W9-CFX-247

ARCTIC ANIMALS

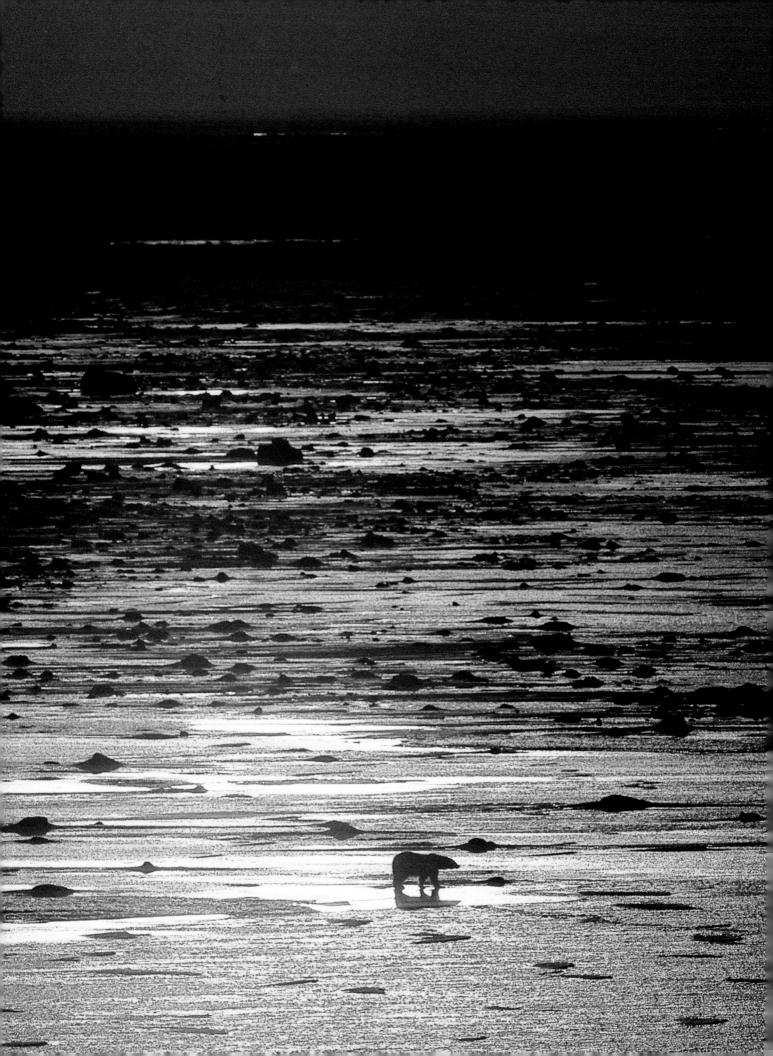

ARCTIC ANIMALS

A CELEBRATION OF SURVIVAL

Copyright © 1986 by Fred Bruemmer

Reprinted 1989

All rights reserved. The use of any part of this publication
reproduced, transmitted in any form or by any means, electronic,
mechanical, photocopying, recording, or otherwise, or stored
in a retrieval system, without the prior consent of the
publisher is an infringement of the copyright law.

ISBN 1-55971-020-9 h c

ISBN 1-55971-021-7 p b

Library of Congress Catalogue Card Number 87-6124

Printed and bound in Hong Kong by Scanner Arts

DESIGN: Brant Cowie/Artplus Ltd.

Published in the United States of America 1987 by

For a free color catalog describing
Northword's line of natural history books
and gifts call 1-800/336-5666.

NorthWord
PRESS, INC

Box 1360 • Minocqua, Wisconsin 54548

PAGE ONE: *Rising like a shaggy giant to watch an approaching bear, a polar bear male raises his front legs to keep
his balance.*

PAGE TWO: *Polar bears at nightfall in early winter upon
the ice of Hudson Bay.*

PAGES FOUR/FIVE: *Hunted intensively by Inuit and nineteenth-century explorers and whalers, muskoxen received
total protection in Canada in 1917. Now they are again
hunted, but on a quota basis.*

PAGES SIX/SEVEN: *Steller sea lion males at sunrise on a
"bachelor beach" in Alaska. While the most powerful males
rule the breeding rookeries, the old, the young, and the feeble
rest on separate beaches.*

PAGE EIGHT: *Walruses like togetherness and lie tightly
packed upon an island beach. When walruses are warm,
blood vessels in their skin expand, and the massive animals
turn from rosy- to brick-red.*

Contents

Preface

THE SUMMER of 1981 was fascinating and typical. First I camped for four weeks in a superb valley, hemmed by multi-hued mountains and glittering glaciers, on northeastern Ellesmere Island, less than 800 miles from the North Pole, together with Dr. R.I.G. Morrison of the Canadian Wildlife Service and his assistant Chris Rimmer. Dr. Morrison is a world authority on bird migrations. Waders are his specialty, particularly such relatively little-known species as Baird's sandpiper and the knot, both of which were common in our valley.

Each day we spent ten to twelve hours watching waders and locating their nests. We banded adults and chicks and studied the growth and dispersal of the young. At night, in our cramped tent, we talked endlessly about waders or read reams of scientific papers about the waders and their ways. Not surprisingly, I learned a lot about waders during that month.

From Ellesmere, I flew to Cunningham Inlet, Somerset Island, where Dr. Thomas G. Smith of the Arctic Biological Station, Department of Fisheries and Oceans, was studying white whales, assisted by two friends of mine, Wybrand Hoek and Gary Sleno. Then my days were spent watching whales (this was my third season with the white whales) and my evenings talking or reading about them.

In September, I flew to northern Banks Island to observe muskoxen during their rutting period. I watched in awe as the great bulls fought each other for control of the herds. My field season ended with a trip in October and November to Cape Churchill on the west coast of Hudson Bay to study polar bear behaviour, my seventh year with these magnificent animals.

In 1982, I was again in Canada's central Arctic; in 1983 in the western Arctic, and in 1984 in Siberia. Thus, for more than twenty years and usually for at least six months of each of those years, I have roamed the Arctic and watched its wildlife in endless wonder and fascination: in Alaska, in all parts of the Canadian North, in Greenland, Siberia, and Lapland, and on Spitsbergen, often alone, camping for weeks and sometimes for months on the tundra or on some remote island, occasionally in the company of scientists who generously shared with me their great fund of knowledge.

Returning to its well-hidden tundra nest, a Baird's sandpiper passes bright-purple saxifrage.

For their help, advice, encouragement, and friendship I am deeply grateful to Dr. A.W. Mansfield, Dr. David E. Sergeant, Dr. Thomas G. Smith, Dr. Edward D. Mitchell, Mr. Wybrand Hoek and Mr. Gary Sleno of the Arctic Biological Station, Department of Fisheries and Oceans; Mr. Brian Beck of the Bedford Institute of Oceanography; Dr. Ian S. McLaren of Dalhousie University, Halifax; Dr. A.H. Macpherson, Dr. Ian Stirling, Dr. David N. Nettleship, Dr. R.I.G. Morrison, Dr. A.J. Gaston, Dr. Austin Reed, Dr. T.W. Barry, Mr. Frank Miller, Mr. Charles Dauphiné and Mr. David Noble of the Canadian Wildlife Service; Dr. Charles Jonkel of the University of Montana; Dr. David R. Gray and Dr. S.D. MacDonald, National Museum of Natural Sciences, Ottawa; Dr. David A. Sherstone and Mr. John D. Ostrick, Inuvik Scientific Resource Centre, Indian and Northern Affairs; Dr. Donald L. Pattie, Northern Alberta Institute of Technology; Dr. Dorothy E. Swales, Macdonald College, Ste. Anne-de-Bellevue; Dr. Norman Barichello and Dr. David H. Mossop, Yukon Department of Renewable Resources; Dr. G. Burton Ayles and Mr. Richard T. Barnes, Freshwater Institute, Fisheries and Oceans; Mr. Brian Davies, International Fund for Animal Welfare; Dr. J.E. Lewis, Atlantic Marine Wildlife Tours Ltd.; Dr. David Cairns, Memorial University, Newfoundland; Mr. James Faro and Mr. Donald Calkins, Alaska Department of Fish and Game; Mr. Lorne Walker, Environmental Department, Mobil Oil Canada Ltd.; Walt Cunningham and Susan Stanford of Alaska; Dr. Christian Vibe of Copenhagen, Denmark; Mr. Len Smith, Tundra Buggy Tours Ltd., Churchill, Manitoba; John and Caroline Kroeger, New York State, companions on many northern trips.

I have been fortunate to make several trips together with two of the world's finest wildlife photographers and very dear friends, Dan Guravich and Stephen Krasemann.

Canada's Polar Continental Shelf Project has assisted me on several occasions and for this I am very grateful. My special thanks go to its director, Dr. George Hobson, to its base managers Eddie Chapman and Barry Haugh, and to the former base manager in Resolute, Fred Alt.

Above all, I am grateful to my wife, Maud, for her love and understanding, and for letting me leave, without plaint, year after year to wander for months in the far reaches of the North.

For entirely different reasons I am deeply, vitally indebted to Dr. Magdi H. Sami and Dr. J. Symes of the Royal Victoria Hospital in Montreal and to Dr. M.D. Rosengarten of the Montreal General Hospital. Without their great knowledge and skill this book would never have been completed.

RIGHT: *Champion migrant of all birds, an arctic tern rests on a lichen-speckled boulder in the Far North. In winter, some arctic terns fly as far south as Antarctica.*

FOLLOWING PAGES: *As jackals follow lions, hungry arctic foxes follow polar bears in winter hoping for leftovers.*

The Region of Darkness

FAR to the north, beneath Polaris, the North Star, and Arktikos, the constellation of the Great Bear, there existed, wrote the Greek historian Herodotus in 430 B.C., a vast land kept in thrall by "an altogether unsupportable cold." This, said the traveller Marco Polo in 1296, "is called the Region of Darkness . . . [where] during most of the winter the sun is invisible," a fearsome, frozen land inhabited by strange creatures such as "bears of a white colour and of prodigious size."

It was the Thule of the Greeks, the Ultima Thule of the Romans, the utmost bound of the world, terrifyingly cold and dreary, home of Boreas, god of the icy north wind, a place of fear and fable and of fabulous animals for which southern man was willing to pay fortunes. Nine hundred years ago, gyrfalcons, the superb white falcons of the Arctic, were traded to China, to central and southern Europe, and even to the Middle East, where they were the pride of its princes. (Traditions linger. Oil-rich Middle East princes still pay fortunes–as much as $65,000 each–for gyrfalcons caught illegally in the Arctic and smuggled to their realms where they are now the ultimate in status symbols.) In mediaeval times, the narwhal's tusk, the straight, tapered tooth of a small arctic whale, was sold in the South as the wonder-working "unicorn's horn," which supposedly could detect and neutralize poison, cure most ailments, and was worth many times its weight in gold.

The Arctic's very aura of mystery and remoteness made its animals immensely desirable and valuable. Kings coveted them for their menageries. Rulers sent them as gifts to other rulers. Ptolemy II, king of ancient Egypt (285-246 B.C.), kept a polar bear in his private zoo at Alexandria, though how he got it the records unfortunately do not tell. When the Norse chief Othere visited King Alfred the Great of England, he told him of his travels to the "furthest North" (really Russia's White Sea) in the year 890, where he had obtained walrus tusks of "great price and excellencie." Mary Tudor, queen of England, sent Tsar Ivan (the Terrible) of Russia "a Male and Female Lions" and received from him "a large and faire Jerfawcon" (gyrfalcon); one of his successors, Tsar Boris Godunov, presented seven "fish teeth" (narwhal tusks) to Shah Abbas the Great of Persia. (This tradition also persists. When President Nixon visited China in 1972, he gave the People's Republic two muskoxen, and his hosts reciprocated with a pair of pandas.)

In the sixteenth century, Europe's perception of the Far North changed abruptly. No longer was the Arctic seen merely as a remote region of darkness,

Snow geese that nested on Banks Island in Canada's western Arctic migrate above the tundra towards the south in fall.

whence came weird beasts to titillate the rich, but as a monstrous, though surely not insuperable, obstacle to the mercantile ambitions of nations. Spain and Portugal had discovered, and pre-empted, the southern routes to India, China, and the Spice Islands of the Far East. Barred from these southern approaches, England and Holland, both burgeoning maritime powers, sought alternate routes to the infinite wealth of the East. Thus began the centuries-long quest for a Northwest Passage and a Northeast Passage.

Both efforts failed in their main objective, to find a commercially viable northern sea route to Asia, for, as Milton remarked in *Paradise Lost*, the explorers ran into "Mountains of ice, that stop the imagined way . . . to the rich Cathaian coast." But these expeditions did result in the discovery of the North's great animal wealth and soon in its widespread and ruthless exploitation.

In 1553, England dispatched the navigator Richard Chancellor to search for the Northeast Passage. He managed to reach only Russia's White Sea, but was heartened when he found upon its shores a narwhal tusk, for "knowing that Unycorns are bredde in the landes of Cathaye," he assumed he was not far from China. Stymied by ice in his efforts to get there, he travelled instead to Moscow at the invitation of Tsar Ivan IV, and out of their negotiations grew the Muscovy Company, which funnelled the fur wealth of Russia's North to England and western Europe. "As for Sables and other rich Furres they be not every mans money: therefore you may send the fewer," the company instructed one of its agents in Russia in 1560. "[But of] martens and mynkes . . . you may send us plentie."

While growing rich in its trade with Russian furs, the Muscovy Company had not forgotten the even greater wealth of the East, and in 1607 it asked Henry Hudson to sail to China via the North Pole. There was then a widely held belief, which persisted well into the nineteenth century, in the existence of an "open polar sea" hemmed by a band of ice. If only this icy barrier could be broached, the argument went, the rest of the trip across the top of the world would be plain sailing. Ice stopped Hudson at 81° north latitude near Spitsbergen but he did report that this region abounded in bowhead whales, "which whale," the Muscovy Company happily noted, "is the best of all sorts": huge, slow, and sheathed in blubber two feet thick. The blubber of one whale could be rendered into twenty tons of valuable oil, which was greatly in demand by the soapmakers of England. Soon hundreds of vessels from many nations were sailing the northern seas in search of whales.

The great arctic hunt for whales and seals and walruses, and for sea otters and sables and foxes, lasted for centuries. In a minor way, it persists to this day. But while the northern seas were exploited and some of their coasts were explored, enormous regions of the North remained virtually unknown. When James W. Tyrrell of the Geological Survey of Canada crossed in 1893 that "great mysterious region of terra incognita commonly known as the Barren Lands," an area

---

Squabbling Steller sea lion females. While adult males fight viciously, females argue and spar a lot but rarely injure each other.

larger than France, "of almost this entire territory less was known than of the remotest districts of 'Darkest Africa.' "

Yet already in Tyrrell's time muskoxen were becoming rare because, he said, the Inuit "pursue a policy of systematic slaughter in quest of the princely robes so much in demand by the fur-traders." Relentlessly hunted, the bow-head whale was failing fast. The great auk had long been extinct. The million-strong flights of Eskimo curlews that once "darkened the sky" had vanished like a northern mirage.

Despite such haunting examples of man's rapacity and its results, nineteenth-century man was essentially optimistic. Only a hundred years ago, the famous English scientist Thomas Henry Huxley could state with utter confidence: "I believe . . . all the great sea fisheries are inexhaustible; that is to say that noth-ing we do seriously affects the number of fish." In 1907, the naturalist Ernest Thompson Seton visited the "Arctic Prairies" and noted, correctly, the sparse and slow growth of northern plants. He nevertheless equated the tundra's food potential with that of Illinois, which then had three million cattle, and con-cluded from this that the caribou on the vast ranges of the North must "number over 30,000,000 and may be double of that" and felt that nothing could ever threaten them. They numbered, in fact, about three million and were soon to begin their drastic decline.

Out of such conceptions and misconceptions grew the modern vision of the North that oscillated wildly between the still-popular notion of a "Frozen Wasteland" and the equally misleading portrait of a benign and "Friendly Arctic." Only in recent decades have we gained a deeper understanding of the Arctic as a realm apart, with plants and animals superbly adapted to its hard conditions, the last great wilderness on earth, vast yet extremely vulnerable.

The land we now call the Arctic was born in an instant of geological time when the giant glaciers of the waning ice age melted and receded. Today's Far North mirrors, to some extent, ice-age conditions a mere 20,000 years ago, when the Arctic reached far to the south, when nearly a third of the earth's land surface was covered by mile-thick ice sheets, when our Cro-Magnon ancestors hunted mammoths in Spain and reindeer in France, when walruses lazed on the beaches of South Carolina and the Bay of Biscay, and mastodons and muskoxen grazed on the periglacial plains of what is now New York State.

A young ringed seal after its first moult. Ringed seals are mainly coastal animals. They prefer the bays and inlets of the North.

ABOVE: *A walrus bull with broken tusks. Small or broken tusks diminish a male's social standing.*

RIGHT: *A massive walrus bull rears up. His long, sharp ivory tusks are used mainly to intimidate rivals.*

The Age of Ice

DURING much of the Pliocene, the eight-million-year-long epoch that preceded the Pleistocene and that saw the gradual emergence of man in Africa, the earth's climate was relatively benign. In the North, deciduous forests and verdant prairies and marshes and meadows reached far towards the North Pole.

About three million years ago, the climate became much colder. In the Far North, yearly snowfall exceeded the annual melt, the snows of yesteryear turned into coarse-grained firn, which, under the pressure of the new layers of snow that were added year after year, changed into glacial ice.

Slowly, inexorably, the growing glaciers crept down the mountains, converged, and coalesced into immense ice sheets. The gradual cooling of the Pliocene culminated in the vast, though cyclical, glaciations of the following epoch, the Pleistocene.

The reasons for these drastic changes in the earth's climate are still obscure. A host of theories has been advanced to account for the ice ages, ranging from fluctuations in the intensity of solar radiation due to the prevalence or absence of sunspots, to cataclysmic volcanic eruptions that spewed such masses of dust into the earth's atmosphere as to diminish seriously the sun's incident energy. Not one of these theories, however, has gained universal acceptance.

Whatever triggered the global chilling, the effects were widespread and affect us still, for we are in the waning era of an ice age, a relatively warm interglacial period in which ice-age conditions prevail only in the arctic and sub-arctic regions. Or we may be, as some scientists fear, near the beginning of a new ice age.

At least four times during the two million years of the Pleistocene, the stupendous ice sheets advanced out of the North only to recede again during interstadial periods that were as warm as our present age and at times even warmer. The last great ice age, called the Wisconsin in North America and the Würm in Europe, began about 120,000 years ago. Once again the icy juggernaut from the North ground southward across the land. At the peak of its expansion 20,000 years ago, ice sheets about 3,000 feet thick in Eurasia and more than 10,000 feet thick in North America covered all of Canada and the northern parts of the United States, much of northwestern Russia, all of Scandinavia, a great portion of northern Germany and Poland, and most of Great Britain and Ireland.

An angry muskox bull paws the ground and rubs his head against a foreleg. If such warnings are ignored, the bull may suddenly charge.

Nearly six million square miles of North America and two million square miles of northern Eurasia lay buried beneath this monstrous carapace of ice. Siberia, however, and much of Alaska were nearly free of ice, and both were then much larger than they are now, for locked within the ice sheets of the North were ten million cubic miles of water that had come from the sea. As a result, the level of the world's oceans was nearly 400 feet lower than it is at present. England and Ireland were part of the European continent; Sri Lanka was attached to India, Japan to the Asian mainland, and in the North a vast land bridge, nearly 1,000 miles wide, the now-vanished land of Beringia, connected America and Asia so that the two continents formed one immense, continuous land mass.

Across the broad land corridor of Beringia, a lake-dotted tundra-like plain, animals drifted freely from continent to continent. From Eurasia to America came bison, moose, mammoth, caribou, muskox, the giant lion, and a host of other species. Migrating the other way were such natives of the Americas as the ancestors of the modern horse and camel, wolves, foxes, and the woodchucks. Puffy-nosed saiga antelopes, now restricted to the steppes of southeastern Russia, ranged from France to Alaska. Mammoths browsed in southern Germany, on the now-drowned plains that stretched nearly 900 miles north of today's Siberia, and in northern California and Wisconsin.

Plants, too, spread around the top of the globe. Today, of the nearly 1,000 species of arctic vascular plants, more than 200 species are circumpolar. And to judge from food remains found in the stomachs of mammoths preserved in Siberia's permafrost ground, the northern plants of 15,000 years ago were similar to those found now in the Arctic. Many arctic mammals—the foxes, weasels and wolverines, the wolf, the beaver, the lynx—are virtually identical in Eurasia and North America. The moose of one continent is the elk of the other; the caribou of the American North is considered to be the same species as the reindeer of Siberia and northern Europe. Of the seventy-five species of birds that breed in the Canadian Arctic, more than half nest also in the Eurasian North.

The Pleistocene was an age of giants, with beavers the size of bears, ground sloths two storeys high that ranged as far north as Alaska, huge camels, and *Bison latifrons*, an early bison so large its horns measured nearly eight feet from tip to tip. It was the age of mastodons, and of mammoths so numerous that half the world's ivory comes from their imperishable tusks.

Preying upon these huge and abundant herbivores were six-foot-long dire wolves, giant lions that once were common in California as well as in Alaska and Siberia, and sabre-toothed tigers with six-inch fangs, larger than any tiger today. And high in the sky, looking for leftovers from gargantuan meals, circled Teratornis, a condor-like carrion feeder, with a wingspan of more than twelve feet, the largest flying bird of all time.

A harp seal female surfaces between ice floes to look for her pup. More than seventy million of these seals have been killed in 200 years of commercial sealing.

The most successful predator of all, though, was man. Cro-Magnon man was, as the Bible says of Esau, "a cunning hunter," and he left us, in stunning polychrome paintings in such caves as at Lascaux in France and Altamira in Spain, a superb record of the ice-age animals he pursued. He invented the spear thrower, a device that, using the lever principle, more than tripled the strength of thrust and hence the distance to which a flint-tipped spear could be thrown with deadly effect and which is used to this day by the Polar Inuit of northwest Greenland. He trapped animals in pitfalls. He chased herds of wild horses over cliffs, or mammoths into swamps for easy slaughter. He collected clams and mussels, and with a leister, a trident-like spear, impaled salmon trapped in stone weirs. Nearly identical leisters were used by Canadian Inuit until a few years ago.

He knew how to fashion eyed needles of bone and, probably using dried animal sinews as thread, made fur clothing from the skins of cold-adapted animals. Thus protected against the arctic cold, he ranged far north and east into game-rich Siberia.

Small bands of hunters and their families, pursuing herds of animals across the northern plains passed, unknowing, from Asia across Beringia into America, at least 40,000 years ago. These were the ancestors of the Indians who eventually spread across the Americas to Tierra del Fuego, at the southern tip.

About 14,000 years ago, the Wisconsin/Würm ice age began to wane. It became warmer, and the ice sheets receded, baring a land riven and rasped by the billion-ton glaciers, ribbed with moraines, and strewn with glacial till and debris. Meltwater rushed in silt-laden torrents towards the sea. The volume of water then carried by the Mississippi was, scientists estimate, at least six times greater than it is now.

At about the same time, many of the great Pleistocene animals began to vanish. About 10,000 years ago, in North America and earlier in Eurasia, most of the ice-age giants became extinct. The change in climate, the northward advance of coniferous forests, and the rise of the melt-fed seas, which flooded the northern plains, may have contributed to their demise. But it is generally believed that it was man, that most cunning and efficient of predators, who exterminated most of them.

One theory holds that man was so new to these pristine northern regions that its animals had no innate fear of him. They were relatively tame and trusting and thus easily hunted to extinction. "What havoc," said Charles Darwin, "the introduction of any new beast of prey must cause in a country, before the instincts of the indigenous inhabitants have become adapted to the stranger's craft and power."

Caught between crafty man and hostile nature, the megafauna of the Pleistocene perished. More than 200 species vanished; only their bones or tusks remain, mute witness to their passing. The explorer Joseph Billings travelled in Siberia

A massive northern fur seal male surrounded by his "harem." Males weigh up to 600 pounds, the sleek females only 80 to 100 pounds.

at the behest of Catherine the Great from 1785 to 1794 and reported: "Mammoth tusks are found about the Siberian rivers and the shores of the Icy Sea, and scattered all over the arctic flats. . . . It appears that the animal is extinct."

A few of the animals lingered. The mastodon may have survived until about 6,000 years ago in the vicinity of the Great Lakes. In Europe, the mighty aurochs, ancestor of our domestic cattle, lived past the Middle Ages. The last died in Poland in 1627. The muskox vanished from Europe about 3,000 years ago, but lasted a thousand years longer in Siberia and until the 1850s in Alaska. Many remained in the Canadian Arctic and Greenland.

And many Pleistocene animals, though not the largest, survived: the bison of North America and its cousin, the European wisent; moose and elk, caribou and reindeer; the wolf and the huge Kodiak bear, as well as a host of smaller animals, squirrels, lemmings, foxes, and voles, which still inhabit the circumpolar taiga and tundra.

The advance and retreat of the Pleistocene ice sheets marked and moulded the face of the North. The glaciers gouged and grooved the land. Myriad lakes filled these depressions, and became home of the North's great waterfowl wealth and breeding place of the clouds of mosquitoes that plague man and beast in summer, but are an important food for many tundra birds. Eskers, the accumulated sediment of sub-glacial streams, meander across the North like misplaced railroad embankments; their sandy soil makes them ideal denning areas for arctic foxes, wolves, and ground squirrels. Erratics, great boulders carried along by the ice sheets and dropped at random all over the North, are favourite perches of owls and hawks and jaegers, and are often streaked with bright orange lichens that thrive on the birds' nitrogen-rich droppings.

When the ice sheets receded, they left a raw and naked land, thinly smeared with silt and clay and strewn with rocks and rubble. Hardy lichens invaded this wasteland, and tiny cushions of mosses. They grew and decayed, creating with infinite slowness a thin layer of earth upon which more advanced species could grow, the vascular plants, the shrubs, and, finally, the forest. The plants that gradually draped the North were the food basis of all land animals, from minute springtails to mighty muskoxen, and, indirectly, of the predators that hunted these herbivores.

Neither the retreat of the ice sheets nor the northward march of plants and animals proceeded at an unfaltering pace. Cold spells, lasting centuries, delayed or even halted the trend, warm periods hastened it. Most "arctic" animals moved northward in step with the climatic zone to which they were best adapted. A few remained behind in favoured areas. When Europeans reached North America, walruses were common on Sable Island in the Atlantic off Nova Scotia, on the latitude of Milan, Italy. (Hunters from Boston killed the last walruses on Sable Island early in the eighteenth century.) White whales, though now sadly diminished to about 500 from the thousands at the time of

Innocence incarnate, with pure-white fur, large limpid eyes, and droopy whiskers, harp seal pups are immensely appealing.

Jacques Cartier, still live in the St. Lawrence River, an ice-age relic population 1,000 miles south of the Arctic's white whales. Conversely, cod occur in several arctic lakes far north of their present range, trapped there during warm climatic interludes of the past.

One such warm period occurred a thousand years ago when the Vikings settled Iceland and colonized from there the southwest coast of Greenland, and Inuit hunted bowhead whales in the far reaches of the North. In the fourteenth century it became colder; it was the beginning of the Little Ice Age. The Inuit died out on Canada's far-northern islands. The Greenland Vikings vanished, victims, presumably, of the worsening climate and of the encroaching Inuit who were culturally superbly cold-adapted. In Europe, Dutch skaters held winter races on ice-covered canals, and Londoners roasted oxen on the frozen Thames.

The Little Ice Age ended about a century ago, and once again it became warmer. In Canada, moose moved north to the very edge of the Arctic Ocean and black bears became common in northernmost Labrador, their new range overlapping that of the polar bear. The growing season of plants lengthened in this century, a fact reflected in the broader growth rings of Scandinavian spruces and pines. Since 1900, a quarter of all European birds have expanded their range northward.

But beginning about ten years ago, the climatic pendulum appears to be swinging again towards cold. Satellite pictures show an increase in the snow and ice cover of the North. Cod, nearly non-existent along Greenland's coast during the Little Ice Age, and then so numerous during this century's warm periods that the cod fishery became, for a while, Greenland's major industry, are retreating again towards the South. The growing season on British farms has declined by about two weeks. Harsher winters, as well as inefficiency, seem to be responsible for poor harvests in recent years in the Soviet Union. The North is cooling, and some scientists fear another ice age is about to begin. And, according to the famous British scientist Sir Fred Hoyle, "when the next ice age comes, it will come quickly . . . [and] will plunge us back into conditions survived by our ancestors–conditions that will be disastrous for our present-day civilization." The gloomy Norse predicted this. In the dreaded "Winter of Ragnarok," their myths foretold, both man and gods would perish in a "world rushing to universal ruin."

> There is fear
> In feeling the cold
> Come to the great world
> And seeing the moon
> –Now new moon, now full moon–
> Follow its old footprints
> In the winter night.   *Inuit poem, recorded by Knud Rasmussen in 1923.*

A female hooded seal defends her newborn pup. The pups grow very fast. Weighing 30 pounds at birth, they balloon to 150 pounds in less than two weeks.

The Hunters and the Hunted

WHEN the ancestors of the Inuit, a people of Arctic Mongoloid stock, crossed from Asia to America about 10,000 years ago, Beringia, the 1,000-mile-wide land bridge, had probably already vanished, like an arctic Atlantis, beneath the rising sea of the fading ice age. But the fifty-seven-mile-wide Bering Strait, which now separates the continents, was a minor obstacle. The Inuit's forbears may have crossed it in skin boats or on winter ice.

I once spent many months on Little Diomede, a small island in the centre of Bering Strait, and the older people among the Inuit who lived there frequently reminisced about the trips to Siberia they had made in their youth. Now the Iron Curtain dividing the Soviet Union from the United States makes such visits impossible, but animals, oblivious to political boundaries, travel as freely between the continents as man once did. Foxes come from Siberia across the winter ice, as do polar bears and occasionally wolves. In spring, some 20,000 sandhill cranes and more than a quarter million snow geese, which have wintered in the United States, fly across the Bering Strait to their breeding grounds in Siberia. And from Asia, flocks of small birds, wagtails, arctic warblers, and wheatears, come across the strait to nest in Alaska.

At the time the Inuit's ancestors crossed Bering Strait, the ancestors of the Indians had already pre-empted most of North America except the Arctic. Barred from southward expansion, these early Inuit turned east and, about 5,000 years ago, occupied the North American Arctic, the harshest, most hostile, and potentially most lethal environment ever inhabited by humans. They spread all the way north to Ellesmere Island and east to East Greenland, tiny pockets of people scattered across the immensity of the North. These Inuit, who in their totality could have easily fitted into a medium-sized football stadium, occupied a land area bigger than Europe. The average population density of the North American Arctic then was about one person to every 250 square miles.

They roamed at will, restricted by neither borders nor exclusive hunting grounds; such concepts were alien to them. "The rights of possession to a locality and its products is originally foreign to the Eskimo mind," noted the anthropologist Kaj Birket-Smith in the 1920s. They lived–they had to live– as Moreau S. Maxwell of Michigan State University has pointed out, in an "adaptive equilibrium" with the game resources of land and sea, and where these resources were stable, their population, too, was remarkably stable.

At the treeline in the North, the vast boreal forest ends and the treeless tundra begins. Hardy spruces and larches, often stunted by wind and cold, are the northernmost trees.

Along the south coast of Baffin Island, for instance, the Inuit, scattered in many camps, numbered about 150 four thousand years ago, about 200 at the time of Christ, 250 when Henry Hudson "discovered" them in 1610, and now, concentrated in one settlement, Lake Harbour, they number about 260.

Not only was the Inuit's land exceedingly cold, hostile, and barren, it was also poor in those raw materials, such as metal and wood, most societies have considered essential. All their land and sea gave them was stone, sod, ice, and snow. Infinitely more important were the materials they obtained from the animals they killed: bone, horn, baleen, antlers, teeth, ivory, furs, skins, sinews, and intestinal tissues. As Dionyse Settle, the Elizabethan chronicler of explorer Martin Frobisher's second expedition to Baffin Island, so shrewdly observed in 1577: "Those beastes, flesh, fishes and fowles, which they kil, they are meate, drinke, apparell, houses, bedding, hose, shooes, thred, saile for their boates . . . and almost all their riches."

To the Inuit the arctic animals meant life. Like other predators they had to live in balance with their prey. If their hunts were too successful and the animals upon which they depended for food declined, the people starved and died until the balance between the hunters and the hunted had been redressed. Before the white man came, said the scientist Ian McTaggart Cowan, "the native people were a dynamic element in the balanced ecosystem."

This total dependence upon animals moulded the Inuit's mental world in ways that may have been similar to those that shaped the mental world of Europe's ice-age hunters. They imbued all animals with souls and felt linked to them by mystic ties of kinship. This animal mystique pervaded their thoughts and dreams, their myths and legends, their actions and their art. They saw animals in the stars and constellations. The Big Dipper was *tuktu*, the caribou; the Pleiades were the "little foxes"; and the Milky Way was made by the "tracks of the great raven." They hoped that wearing animal amulets would, through sympathetic magic, give them the attributes of animals: a wolf's paw to become a hunter as enduring and hardy as the wolf; a ptarmigan leg to blend into the landscape like the cryptically coloured ptarmigan; a bit of ermine fur, to be as quick and agile as the ermine.

Man did not have pre-eminence over animals, nor did he belong to a separate realm. As the Inuk Nalungiuq of the Netsilingmiut explained to an ethnologist in 1924: "In the very earliest time . . . a person could become an animal if he wanted to and an animal could become a human being. Sometimes they were people and sometimes animals, and there was no difference. All spoke the same language." This belief is reflected in many legends. In one, Kiviuq, the mythical hero of many Inuit tales, saw a lovely fox who shed her skin and laid it on a rock to dry. When the vixen entered her den, Kiviuq stole the skin. The fox returned, saw that the man had taken her skin, and pleaded with him to return it. "Only if you marry me," said Kiviuq. The vixen agreed, received her skin, they married and lived, presumably, happily ever after.

A young Steller sea lion male in the swirling surf of an Alaskan island. Now protected through much of their range, these sea lions, the largest in the world, now number about 250,000 to 300,000.

The theme of this story is universal and archetypal; it occurs in thousands of legends and fables around the world that tell of werewolves, or of maidens who changed into swans, as in the story of Swan Lake. It may hearken back to an age when all humans were primarily hunters who identified closely with the animals that were their life. Until quite recently many Scottish fishermen regarded grey seals with superstitious dread, for they believed the ancient tales that they were really "selchies," men and maids turned by enchantment into seals, and they called them "the folk of the sea."

Since the Inuit believed that animals had souls and regarded them, to some extent, as brothers under the skin, an obvious ethical problem arose when they had to kill them in order to eat. The people were quite aware of this quandary. "Life's greatest danger lies in the fact that man's food consists entirely of souls," an Igloolik Inuk told the ethnologist Knud Rasmussen. "We fear the souls . . . of the animals we have killed." The Inuit tried to circumvent this problem with the twin concepts of reincarnation and respect.

After death, the people believed, animals were reborn, and provided an animal in life and in death had been treated with the utmost respect by a hunter, it would allow itself to be killed again in its next incarnation. As a result of such beliefs, a multitude of rites and taboos had to be observed to propitiate the souls of dead animals. Whales and seals, when hauled from the sea, were thought to be thirsty, so water was poured into their mouths. Ekalun, an old Inuk of Bathurst Inlet, with whom I lived for half a year, never failed to thank a seal profusely for permitting itself to be killed and thus providing us with food. And when hunting was unsuccessful, the Inuit consulted the *angakkut*, their shamans, to discover in which way they might have offended the animals.

Out of such beliefs grew a certain ethic and harmony and many eminently sensible rules. Wasteful hunting was regarded as sinful, for an animal's soul would be deeply offended if its body was not used as food. It was taboo to hunt walruses on *ooglit*, their hauling-out islands, for if they were disturbed there too often, they might leave for regions unknown. It was forbidden to camp on *pikyoolak*, the holms on which eider ducks nest, for then the brooding ducks might depart and there would be no harvest of eider eggs in future years.

Then came the whalers who killed for gain and who often took from a sixty-ton whale only the valuable baleen, leaving the giant carcass to rot; the traders who needed furs; the missionaries who scoffed at shamans and taboos; the guns that made hunting infinitely easier. All these caused the ancient beliefs to lose ground; the mystic bond between animals and arctic man was broken and was superseded by alien values and ways. "Now that we have firearms it is almost as if we no longer need shamans and taboo," said Ikinilik of the Utkuhikhalingmiut in 1924.

With the dissolution of the intense mystical and spiritual ties that once bound man to the animals of the North, an ancient way of life ceased. Many

Narwhal males, the ivory-tusked "sea unicorns" of the arctic seas, whose "horns" in mediaeval times were worth many times their weight in gold.

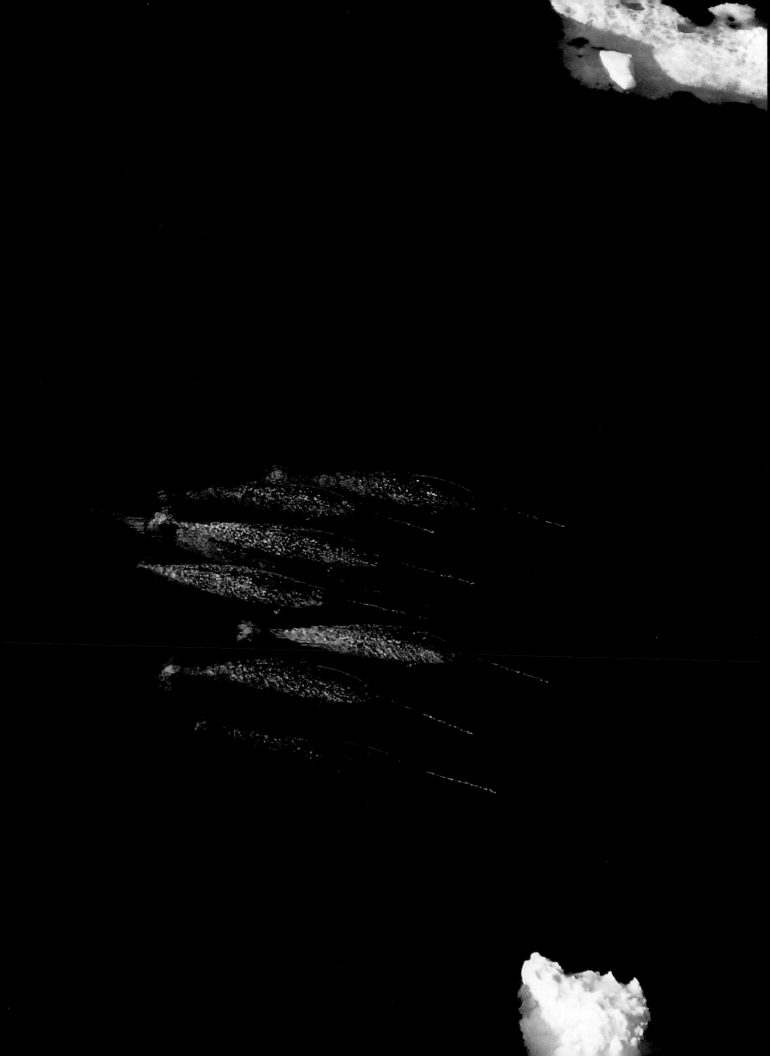

older Inuit mourn its passing. Reminiscing about the changes he had seen, Koonoo Muckpaloo of Arctic Bay on northern Baffin Island said in 1975: "It seems it was such a short time ago that we were still living in our own way and today when you look around, it is all dying out. . . . For myself, I am sad that the Eskimo way has gone."

Oh warmth of summer sweeping o'er the land!
Not a breath of wind,
Not a cloud,
And among the mountains
The grazing caribou
The dear caribou
In the blue distance!

I lie upon the ground sobbing with joy.

Song of the Polar Inuit, recorded by Knud Rasmussen.

RIGHT ABOVE: *A caribou runs with magnificently long-paced strides across the tundra. When pursued by wolves or man, caribou break into a fast but ungainly gallop.*

RIGHT BELOW: *A caribou herd, migrating south in fall, splashes through the shallow water of a tundra lake in northern Canada.*

The Land of Dearth and Plenty

WHEN Europeans first came to the Arctic, they were awed by its immensity, appalled by the harshness of its climate and the seeming barrenness of the land, and occasionally enthralled by the wealth of its wildlife. The taiga, the primarily coniferous circumpolar boreal forest belt, and the treeless tundra north of it cover together nearly a quarter of the world's land surface. The tundra can be vibrant with life in summer and hauntingly dreary and desolate in winter. "A monotonous snow-covered waste," wrote the nineteenth-century English traveller Warburton Pike after crossing Canada's tundra in winter. "A deathly stillness hangs over all, and the oppressive loneliness weighs upon the spectator until he is glad to shout aloud to break the awful spell of solitude."

The coldest areas of the North are not, as one might expect, in the high Arctic but well to the south within the forest belt at Verkhoyansk, Siberia ($-94°$ F.), and near the village of Oymyakon, where in the winter of 1964 the temperature dropped to a record $-96°$ F. Alaska's coldest spot is Prospect Creek, just north of the Arctic Circle, where the lowest recorded temperature was $-79.8°$ F. in January 1971. But these regions of extreme cold have relatively long and warm summers, with temperatures as high as 98° F. at Verkhoyansk. And in winter when it is so cold that human breath freezes with a crackling sound which Siberians rather poetically call "the whispering of the stars," the air is utterly still and very dry.

If winters in the Far North are not as extremely cold as within the forest belt, they more than make up for it by being exceedingly long. In the Thule region of northwest Greenland, the average temperature of only one month, July, is above the freezing point. The sea is covered by ice for nine months of the year. The sun sets in October and does not rise again above the horizon until late February. Winter temperatures often plunge to $-40°$ F. and even $-50°$ F. and stay there for days and occasionally weeks (the average temperature for both January and February is $-29°$ F.), and storms are frequent, long, and lethal. (Yet the Polar Inuit who live in the Thule region call it *nunassiaq*, the beautiful land!)

The sub-arctic maritime regions are much more clement but are haunted by storms and fogs. The Aleutian Islands, according to the United States Coast Pilot, have "the most unpredictable [weather] in the world. Winds of up to 90 miles an hour are commonplace, bad weather and storms may be expected at any time of the year. . . . Clear days are practically non-existent." The islands,

---

Surrounded by an aureole of spray, a Kodiak bear shakes water from its shaggy fur.

TOP: *When salmon swim upstream to spawn in Alaska's rivers, giant Kodiak bears wait for them and catch them with a speed and agility amazing in animals so huge and bulky.*

ABOVE: *A moulting harbour seal tries to scratch a hard-to-reach itchy spot.*

RIGHT: *Thick-billed murres lay their large, pear-shaped eggs on the narrow ledges of soaring cliffs, safe from most predators. Parents take turns incubating their single egg and in sheltering and feeding the chick.*

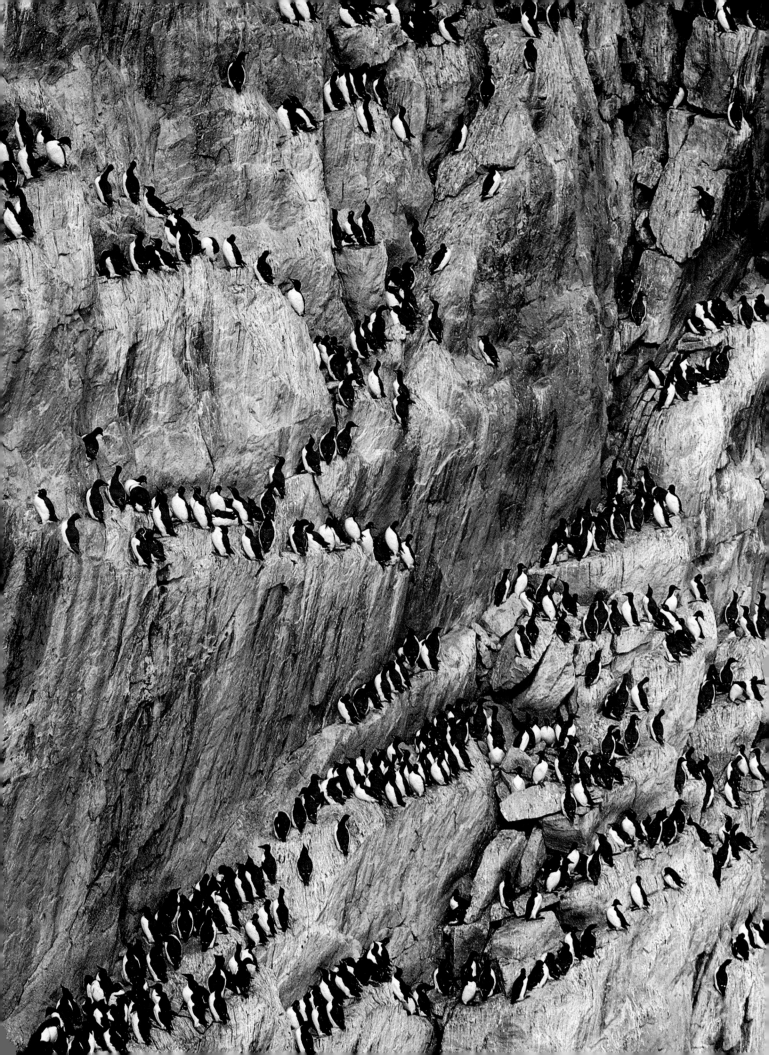

46

wrote the Russian bishop Veniaminov, the "Apostle of the Aleuts," are "the tsardom of winds." He spent ten years there (1824-1834), and during that entire time, he said, "there has not been one single day . . . wholly free of wind." The Pribilof Islands in the Bering Sea have an average of twenty-two clear days a year, forty-two days of heavy fog, eighteen days of snow and sleet, and two hundred days of rain. Attu, the westernmost island of the Aleutian chain, has on the average only seven clear days a year.

Despite such miserable weather, the arctic land and, above all, the arctic and sub-arctic seas were once immensely rich in animal life. In 1890, the famous Russian writer Anton Chekhov spent many months on Sakhalin, an island off Siberia, with weather, to judge from his descriptions, nearly as damp and dismal as that of the Aleutians. But in spring when the herring came into shoal water to spawn, Chekhov wrote, "the sea appears to be boiling over," and the fish were followed by "gulls, albatrosses . . . and herds of sea lions. The scene is magnificent!" In late July the salmon came, and "the surface of the river seems to be seething. The water has a fishy taste, the oars are jammed, the blades propel the obstructing fish into the air." The number of whales near the island "was so great that [the explorer Adam von] Krusenstern's ship was encircled by them and it was only with extreme caution that they could reach" the shore of Sakhalin. Bears were common on Sakhalin, as were sables, foxes, and wolves; but already in Chekhov's time its tigers were nearly extinct.

Out of the North where they nested, clouds of birds came to the South each fall. Snow geese swept over the land like a "snow storm," wrote William Wood of Massachusetts in 1629. "Sometimes there will be two or three thousand in a flocke, and continue six weeks." Until the early 1800s, wild geese and ducks were New England's cheapest food, and slave owners who hired out their slaves stipulated in the contracts that they should not be fed ducks or geese more than twice a week. Eskimo curlews arrived from their breeding grounds on the arctic tundra in swarms that "darkened the sky," with a noise "like the wind whistling through the ropes of a thousand-ton vessel." They were so fat that New Englanders called them "dough birds," and so tame and trusting they could be killed with sticks. They sold for six cents apiece on the Boston markets.

When the explorer Samuel Hearne crossed the "Barren Lands" of Canada in 1771, he often saw "many herds [of muskoxen] in the course of a day's walk, and some of those herds did not contain less than eighty or an hundred head." In 1893, James W. Tyrrell of the Geological Survey of Canada travelled across the Barrens and near the upper Dubawnt River. He commented: "The valleys and hillsides for miles appeared to be moving masses of reindeer [caribou]. To estimate their number would be impossible. They could only be reckoned in acres or square miles." The American explorer Isaac Israel Hayes, searching in 1861 for that elusive "open polar sea," saw walruses near the coast of north-

RIGHT ABOVE: *Fluke and head raised out of the water, an itchy beluga scratches its belly in the shoal water of a high arctic bay.*

RIGHT BELOW: *Like gleaming white torpedoes, belugas glide through the clear, greenish water of a high arctic inlet where these white whales congregate each summer.*

west Greenland and remarked that "they extended as far as the eye could reach, almost every piece of ice being covered." It required a stupendous benthic wealth to support such enormous herds of walruses. The total population of Atlantic and Pacific walruses in the mid-nineteeth century probably exceeded 300,000. If all were feeding, they would eat roughly one billion clams a day.

In late July of 1819, after battling for weeks with heavy ice in Baffin Bay, William Edward Parry of Britain's Royal Navy, commanding the ships *Hecla* and *Griper*, emerged into the relatively ice-free Lancaster Sound, the gateway, he hoped, to the Northwest Passage. He was amazed and delighted, for he had arrived in an arctic Eden, the "headquarters of the whales." Sea mammals abounded. Polar bears ambled across the ice floes, and the sailors tried to lasso them. "Sea horses" (walruses) lay "huddled together like pigs" and were "stupidly tame." Giant bowhead whales lolled lazily in the dark sea. On July 30, they saw eighty-nine. Narwhal, the ivory-tusked whales of the Far North, were "very numerous," and white whales "were swimming about the ships in great numbers" with "a shrill, ringing sound, not unlike that of musical glasses badly played."

Near Loks Land, at the entrance to Baffin Island's Frobisher Bay, the explorer Charles Francis Hall saw in the summer of 1861 so many eider ducks "the water and the air were black with them." Hall and a group of whalers rowed to the islands where the eiders nested. They returned in a few hours, "both boats laden with . . . one hundred dozen eggs." Later, together with Inuit, Hall travelled to the head of Frobisher Bay where "the waters . . . were teeming with animal life" and the rivers were "alive with salmon [char]." Awed by such profusion, Hall exclaimed: "Indeed we are in a land and by rivers of plenty."

Such reports of a bounteous Arctic, while true, are somewhat misleading. Hall, in this "land . . . of plenty," suffered near-starvation on several occasions; many arctic explorers starved to death, and most others only avoided that fate because their ships were amply stocked with food from the South. Fear of famine haunted most Inuit groups; tales of hunger and starvation run like a *leitmotiv* through their oral history. Among the Netsilingmiut, Rasmussen recorded in the early 1920s, even the most tireless and proficient hunters rarely succeeded in killing more than seventy caribou a year and fifty to sixty seals, barely enough to feed their families and sled dogs.

Many years ago, I crossed the ice cap of Spitsbergen with a British expedition, and during five weeks the only animal life we saw were three ivory gulls and several fulmars. When George Hakungak, an Inuk from Bathurst Inlet, and I sledged in March across the tundra to pick up caribou meat cached inland the previous fall, it was an eerie trip, across a hushed, white-shrouded land nearly devoid of life. In eight days of near-constant travel, we saw four ravens, a few ptarmigan, and one dead lemming. "We seem to glide from nowhere to nowhere through a featureless, lifeless land," I wrote one night in my diary.

Massed walrus bulls on Alaska's Round Island. While most Pacific walruses migrate north in spring to the food-rich Chukchi Sea, a few thousand males laze away the summer months on Round Island and in the nearby sea.

Even more daunting and desolate is the ice that covers the polar sea all the way to *Kalalerssuak*, the big navel (of the earth), as the Polar Inuit call the North Pole. They have another, more ominous name for it: *Kingmersoriartorfigssuak*, the place where one eats only dogs.

This paradox of a north that some found so rich in wildlife and others so utterly desolate is due in part to the sheer immensity of the Arctic and to the seasonal and local abundance of many northern animals. Land animals, generally, are widely scattered to utilize most efficiently the sparse and slow-growing vegetation of the North. Canada's caribou once numbered about three million, but their combined summer and winter ranges exceeded 700,000 square miles, an area about fourteen times the size of England. In winter, dispersed in small groups, most caribou browsed on lichen within the boreal forest belt. In spring they moved north, following ancient migration routes, trickles of animals that met and merged into a mighty stream of life flowing across the tundra so that "one can see neither the beginning of them nor the end–the whole earth seems to be moving," as an Inuk told Rasmussen in 1923. But, as the biologist George Calef has pointed out, "those who witness these annual spectacles are often unaware that they are seeing virtually all the caribou from an immense area of country."

Polar bears number at present about 15,000, and their realm is immense, more than five million square miles of circumpolar land and frozen sea. Mathematically, this works out to one polar bear for every 333 square miles. In reality, of course, polar bears are fairly abundant in a few favoured regions and rare in much of the Arctic. Dutch seamen wintering on Spitsbergen in 1623 saw polar "bears going in troops, like cattle in the Netherlands." In 1864, in northern Baffin Bay, the British whaling ship *Constantina* came upon a "dead whale covered with [polar] bears. The ships's crew shot . . . sixty." And every fall as many as 600 polar bears mass along a 100-mile stretch of Hudson Bay coast, between the Nelson and the Churchill rivers, waiting for ice to cover the bay so that they can go out upon it and hunt seals again. Such agglomerations are the exception. In most regions of the North, polar bears are so rare the majority of Inuit have never seen one.

May, in the dialect of the Polar Inuit of northwest Greenland, is called *agpaliarssuit tikiarfiat*, the dovekies return, and August is *ivnanit aorsarnialertarfiat*, the month when the young birds fly to the south. For three and a half months, the soaring scree slopes where the dovekies nest are speckled with millions of these starling-sized, chubby seabirds, and the air throbs with their chirps and cries. Then, quite abruptly, at the end of August, the young are fledged, the dovekies leave, and the great slopes, which briefly pulsated with millionfold life, are still and abandoned.

The Inuit of Little Diomede Island in the middle of Bering Strait call June "the month when people do not sleep." During that month, the walruses,

---

A male narwhal on a beach of northern Baffin Island. In mediaeval times, narwhal tusks were sold in Europe as wonder-working unicorn horns.

about 150,000 now, that have wintered in the Bering Sea migrate north to the Chukchi Sea, all funnelling through Bering Strait, and the Inuit hunt day and night, because once this mighty throng has passed no walruses will remain in Bering Strait and the great hunt will be over for another year.

Some arctic animals are immensely numerous. The dovekies that breed only in the farthest North, on northwest and northeast Greenland, on remote Jan Mayen Island in the Greenland Sea, on Spitsbergen, and on a few island groups north of Siberia, number more than eighty million. The least auklets of the Bering Sea may exceed a hundred million. But while a few northern species are stunningly abundant, the number of animal species inhabiting the Arctic is very small compared to the exuberant profusion of species found in more temperate regions of the world. In one day of 1832, on the outskirts of Rio de Janeiro, Charles Darwin, although he was "not attending particularly to the Coleoptera," collected sixty-eight species of beetles. But all of Greenland, the world's largest island, has only twenty-six species of beetles.

Of the world's 3,200 mammal species, about forty live in the Arctic, and nearly half of those are sea mammals. There are about 8,600 species of birds in the world. A small, tropical country like Ecuador is home to 1,500 species. Yet in the immensity of the Arctic only about 100 bird species breed, and most are migrants who come north during the short but food-rich summer, raise their broods, and flee to the South at the approach of winter. Less than ten bird species live year-round in the Arctic. Of the world's 750,000 insect species, less than 1,000 occur in the North; of the roughly 30,000 species of fish, less than 100 live in the northern seas, lakes, and rivers. No reptiles exist in the Arctic, and although several species of toads and frogs inhabit the boreal forest belt, no amphibian ventures north of the treeline. It may be poor solace to northern travellers, harried in summer by clouds of mosquitoes so dense that breathing becomes difficult and exposed skin is quickly covered by glittering legions busily drilling for blood, but of the several thousand mosquito species in the world, only four exist in the Arctic.

The harshness of the arctic climate, the poverty of its soil, the exceedingly long and dark winters are all inimical to life. But nature, as the philosopher Spinoza remarked, abhors a vaccuum, and although the opportunities were few and fraught with hardship, some plant and animal species adapted to life in the Arctic, their ability to survive cold and often marginal food conditions honed to utmost perfection.

The arctic seas and, to a lesser degree, the arctic lakes are rich in nutrients and consequently in phytoplankton, zooplankton, and in fishes and sea mammals. The arctic land is infinitely poorer; its soil is usually oligotrophic, nutrient-deficient, thin, and often acid. It is arid on slopes, for precipitation is minimal in much of the Arctic, yet waterlogged on valley floors where concrete-hard, impermeable permafrost prevents meltwater from sinking into the ground. In addition to being forced to subsist upon such niggardly, hostile soils, the plants' season of possible growth is extremely short; forty days or less in the high Arctic, and two to three months in the low Arctic. As a result, annual plant productivity in the North, in terms of weight per unit area, is only one to five per cent of that of an equal area in temperate regions.

One of the muskox's favourite foods is arctic willow, a stunted shrub that attains bush-size in sheltered valleys or creeps prostrate across the open, wind-swept tundra. The willow branches the muskox munches so contentedly are only finger-thick. But it may have taken them 200 years to attain that size, and two centuries will have to pass before another willow that size replaces the one now eaten. Caribou moss, really an extremely common, circumpolar lichen, is an important food of caribou and, in Eurasia (where it is called reindeer moss), of reindeer. Few other animals utilize this abundant plant–they cannot tolerate the acid it contains. But caribou eat it without ill effect and thrive on it, for it is rich in carbohydrates. If they browse on this lichen alone, they require about twelve pounds per animal each day. But caribou moss regenerates slowly, growing at an average rate of one-sixteenth of an inch per year, and twenty-five to fifty years may elapse before a close-cropped patch of ground is covered again with a thick carpet of caribou moss.

This sparse and slow growth of arctic plants limits the number of herbivores an area can support, as well as the number of predators, who depend upon them for food. Caribou increase their food supply by extending their range through extensive migrations. Muskoxen wander widely, for each muskox needs every month the forage produced by several acres of arctic land. Most birds spend only summers in the North. Peaks in lemming populations are followed by periods of drastic decline. And while the main reason all arctic animals are warmly wrapped in fur, or fat, or feathers is to shield them from the lethal cold, this superb protection also diminishes the quantity of food they need. Poorer insulation would require a higher rate of metabolism, and additional food as fuel to counteract the cold.

Keeping Warm in an Icy Land

ALL animals produce heat by releasing the energy stored in their food, particularly in the lipids, proteins, and carbohydrates. The normal body temperature of mammals is around 100° F. (100.8° F. in polar bears, 101.2° F. in muskoxen, 98.6° F. in humans, 98.4° F. in arctic ground squirrels) and about 104° F. in birds. These temperatures are constant and vital. When the air temperature is low enough to sap an animal's body heat faster than it can be generated, its core temperature is lowered and it is immediately in deadly trouble. Naked, hairless man, the least adapted of arctic animals, exposed to temperatures of −40° F. and winds of thirty miles per hour–conditions common in the Arctic–dies in about fifteen minutes. The Inuit avoided this fate by dressing in the skins of the Arctic's superbly cold-adapted animals.

Animals have two ways of coping with cold: they can increase their metabolic rate (their internal heat production) by eating large amounts of food, and they can prevent loss of body heat through insulation. Some northern animals are voracious. A seventy-pound sea otter eats daily about fifteen pounds of sea urchins, mollusks, and fish. A polar bear, given the chance, can devour 100 pounds of high-calorie blubber at one meal. Wolves, after a successful hunt, consume a quarter of their body weight in food. Ptarmigan are forever busy snipping buds and shoots; in winter, to keep alive, they require a daily food intake roughly equal to a fifth their body weight. But there is obviously a limit to how much food animals can consume to stoke their body's furnace, and for survival in their icy realm they rely more on heat conservation than on heat production. Their insulation–fat, fur, or feathers–shields them from the cold and enables them to live in the Arctic.

To reduce body surface and minimize heat loss, most arctic animals are rounded and compact. The mighty muskox has small, densely furred ears, short stocky limbs, a massive, chunky body, and a four-inch tail. Its only naked spot is a small bare patch on its nose. Beneath its ample cloak of coarse guard hairs, which are so long they hang nearly to the ground, the muskox is swathed in a thick layer of extremely dense, fine wool called *qiviut* by the Inuit. This voluminous, double-layered fur enables muskoxen to endure, with little loss of body warmth, winds of sixty miles per hour and temperatures of −40° F.

When fleeing or charging an enemy, or when a bull pursues a rival, muskoxen can run with amazing speed even over extremely rough terrain. At all other times, they appear phlegmatic and move slowly and with ponderous gravity.

Densely feathered legs and feet help to keep ptarmigan warm and act like feathery snowshoes.

They eat for a while, then rest and ruminate, wander a bit farther and eat again. In this slow-motion existence, muskoxen expend a minimum of energy and thus curtail their food requirements. In summer, muskoxen drink from melt pools, lakes, and brooks and feed on fresh grasses and sedges; their dung is moist and mucky. Their winter forage is dry, and they consume as little snow as possible because energy is required to convert the Arctic's super-cooled snow into water; their winter droppings consist of dark, dry pellets. So efficient is the muskox at preserving energy, it requires, according to a scientific report, only a sixth of the food that cattle of similar size would need. This frugal use of the Arctic's meagre pasture enables muskoxen to live in regions as hostile as Peary Land in northernmost Greenland, where only one inch of precipitation falls during an entire year, and less than 100 plant species eke out a marginal existence.

To prevent excessive heat loss from its long legs, the caribou maintains two internal temperatures. Its body temperature is near 105° F., while that of its legs is more than fifty degrees cooler. Its veins and arteries are closely aligned, like cables in a conduit, so that the outflowing arterial blood transmits its warmth to the chilled venous blood returning from the limbs. Constriction of blood vessels in the extremities permits a flow of blood just sufficient to keep the legs from suffering frost damage, ensuring at the same time that little of that precious body heat is lost to the icy ambient air.

Caribou hairs are club-shaped, thicker at their tips than at their base, and form a densely packed outer layer with myriad tiny air spaces near the skin and within the fine, curly underwool. In addition, the long guard hairs are filled with air cells. This deep-pile coat is so warm it renders caribou virtually impervious to the worst arctic weather.

If caribou are so well adapted to their arctic environment–some scientists call them chionophiles (snow-lovers)–polar bears are pagophilic. They love ice and spend much of their lives upon it, hunting seals, their favourite and most important prey. Polar bears, too, are compactly built, have small, rounded, thickly furred ears, tails so tiny they are barely visible, and short massive legs with heavily haired, fur-fringed paws. A polar bear attacking a rival or an enemy or pouncing upon a seal can move with lightning speed and agility. Normally, though, polar bears shuffle placidly along, warmly wrapped in thick, two-layered fur, and expend as little energy as possible.

Recent studies of polar bear hairs, conducted with scanning electron microscopes, have revealed that they have an exceedingly complex structure and a core that is nearly hollow. In the summer of 1978, officials at the San Diego zoo were startled when some of their polar bears turned green. In California's warm climate, green algae had established themselves within the bears' hollow hairs.

Polar bear hair and the white, woolly lanugo, or natal fur, of harp seal pups are both capable of transmitting solar energy, including that in the ultraviolet

---

Born thickly wrapped in dense, dark, curly wool, muskox calves can easily withstand temperatures of −30° F. They nurse frequently and begin to nibble grasses and willows within a few days of birth.

ABOVE: *Harp seal mother and pup. The pups are born on the pack ice, drink their mothers' fat-rich milk, and double or triple their birth weight in two weeks.*

RIGHT: *Intensely gregarious, the starling-sized dovekies nest in millions on high-arctic scree slopes. In northwestern Greenland Polar Inuit catch the low-flying birds with long-handled nets.*

range, to the animal's skin, where it is absorbed as heat. Thus, even on extremely cold days, the animals absorb solar heat while the greenhouse effect of their fur prevents outward loss of body warmth by radiation. The Polar Inuit of northwest Greenland wear to this day *nanut*, trousers made of polar bear skin, and they are probably the warmest pants in the world. And, a world away, the giant DuPont company recently developed a polyester fibre, Hollobond, that copies the hollow-core structure of the polar bear's hair.

Only one northern mammal, the arctic ground squirrel, avoids the rigours of winter through hibernation. The Barren Ground grizzly also dens up in winter, wrapped in fat up to four inches thick, but he is merely dormant. His heart rate and breathing slow down but he does not sink into the deep torpor of true hibernation; his body temperature remains close to normal as he sleeps away the long winter, living off the energy stored in his fat reserves. The ground squirrel hibernates for eight months.

During summer, the ground squirrels are frantically busy. They mate, raise their young, and feed so persistently and voraciously that from June to September they more than double their weight and look like fat furry sausages with feet. In addition to accumulating this thick padding of fat, the ground squirrels collect food and carry it in bulging cheek pouches to their winter burrows.

These winter dens are of a very special design. They are built in dry, sandy soil, in an esker, for instance, or a brook bank, and in an area that will be covered in winter by a thick layer of insulating snow. The tunnel leading to the hibernation chamber slopes upwards. Since warm air rises, the small amount of warmth generated by the squirrel will not seep out of its winter nest. The sleeping chamber is thickly padded with dry grass, and by September an energetic squirrel may have stocked its winter home with four pounds of assorted foods.

When the chill winds of late September sweep over the land and ice glazes the tundra ponds, the ground squirrels retire from the hostile upper earth. For a while they nibble at their hoard of food, then curl up tightly in their thick grass nests to minimize exposed body surface, and fall into a deep sleep, their body motors barely idling. Their temperature drops from a normal 98.4° F. to near freezing; the heartbeat slows; the squirrels breathe only once or twice each minute, and for eight months, subsisting as sparingly as possible upon their fat reserves, they hover in that mysterious borderland between life and death.

While nature can easily protect the larger arctic animals from winter's deadly chill with bulky fur and ample fat, this solution to survival cannot be applied to the small mammals of the North, such as the voles and lemmings, for a fur long and dense enough to shield them from the cold would also render them immobile. So instead of an impossibly thick fur, snow is their winter blanket. The myriad air cells within the snow make it an excellent insulator. The temperature beneath three feet of snow can be seventy degrees warmer

Arctic ground squirrels have many enemies. When feeding, they frequently stand upright to look for approaching foes and bolt into their holes the moment they see one.

than the air temperature above it. And this subnivean temperature is nearly constant. While temperatures above the snow may soar or drop by thirty or forty degrees, those beneath the snow layer fluctuate only by two or three degrees.

At the approach of winter, lemmings build globular grass nests beneath the snow. One lemming species grows special "snow shovels": its two middle claws become enlarged and lengthen into broad, horny, efficient digging tools. With these, the lemmings excavate a network of runways beneath the sheltering snow to reach the food they require.

Some birds, too, seek the protection of snow. In the evening, after feeding busily all day, running from willow bush to willow bush on densely feathered feet, ptarmigan rise with clattering wings, soar upwards in a parabolic arc, plunge suddenly into a patch of soft snow and burrow deeply into it to benefit from its insulation during bitterly cold nights. By flying into the snow, they avoid leaving telltale tracks that could guide a fox or wolverine to their sleeping places.

Mammals depend on guard hairs and wool to keep them warm, birds on feathers and down. While an extremely fine wool like the muskox's *qiviut* (one pound of which can be spun into a forty-strand thread twenty-five miles long) is a superb insulator, down is even better, for one gram of down has a total fibre length of thirty-two miles. Down, according to California biologist John Dillon, is "the finest insulating material known." It has "the ability to trap a maximum of air with the least weight of material, and the greatest flexibility of any substance known–natural or man-made."

The layer of down near the bird's skin that traps and holds body-warmed air is shielded by hard, vaned contour feathers, which most birds, and all seabirds, oil assiduously while preening to keep them waterproof. The black and white plumage of many northern seabirds, such as the murres and dovekies, is also an adaptation to cold. The white feathers of breast and belly are filled with air cells and form a non-conductive cushion that protects the birds from the sapping cold of arctic water. The dark back feathers are filled with melanin granules, which absorb and conduct solar heat.

Like caribou, arctic birds have a counter-current circulatory system in their legs. Their body temperature is 104° F., while that of their legs and feet is close to the freezing point. A gull can stand on ice for hours, and a murre's feet are constantly immersed in the icy sea, yet neither loses much body heat from its palmate, naked feet, for only enough blood reaches them to keep them from freezing, and the cooled venous blood returning from the extremities is warmed by its proximity to the blood vessels carrying warm, arterial blood outward.

The temperature of sea water does not fall below 28° F. but heat conduction in water is about twenty-five times greater than in air. Most sea mammals are protected from this sapping chill, which would kill a human in about ten minutes, by blubber, two to three inches thick on a seal and nearly two feet thick on a bowhead whale. The blubber is both insulation and energy reserve, an immensely efficient non-conductive shell that prevents heat loss from the body's warm core. In the frigid sea, the sea mammals' peripheral vascular system is constricted, and blood flow to blubber, skin, and flippers is kept to an absolute minimum.

Two sea mammals have developed a different method of staying warm in the cold seas of the North: a fur so dense, water can never touch their skin. A sea otter is covered by 800 million fur fibres, which trap warm air next to its skin, and the northern fur seal's body carries 300,000 hairs to the square inch. Both coats provide magnificent insulation, provided they are clean. A slick of oil that mats the fur, or food debris that soils it can destroy the efficiency of this furry cloak. Both sea otters and fur seals groom themselves with the greatest care to keep their fur neat, fluffy, and air-filled.

To bitter cold, to meagre fare, to icy seas and searing storms, the animals of the Arctic are superbly adapted. Cold is the normal condition of their realm and they can cope with it. But abnormal weather can be fatal for them. When winter comes but snow delays, voles and lemmings, bereft of their protective snow blanket, often freeze to death.

In the 1880s on the Belcher Islands in eastern Hudson Bay there was a winter with heavy snowfalls. In March it became mild and it rained, turning the snow into waterlogged mush. Suddenly the wind veered to the north, the temperature dropped abruptly, the wet snow froze, and the islands were enveloped for weeks in a hard, glittering carapace of ice. In vain did the caribou scrape and scratch to get at the vegetation. Not even their hard-edged hoofs could break this icy mantle, and all caribou on these islands perished. Foehn winds in winter are Greenland's curse. In the 1930s, these warm winds melted the snow in East Greenland and then a return of severe frost glazed the land. All caribou died out in this region, as did the wolves that depended upon them for food.

Weather patterns in the Canadian Arctic tend to be more stable than in Greenland. But occasionally disaster strikes. In the fall and early winter of 1973, heavy falls of wet snow followed by severe frost duplicated the East Greenland disaster. The Peary caribou of the Far North died in droves, and on some of the high arctic islands more than 80 per cent of all muskoxen perished.

Each spring, about a million eider ducks come to Alaska's North Slope and feed in leads of open water not far from shore. A few years ago, heavy frost and northerly winds in spring sealed the leads, and more than 100,000 eiders died. A similar catastrophe befell the murres of Lancaster Sound in 1978. Normally murre chicks jump, before they can fly, from the 1,000-foot sheer cliffs where they are born into the sea below and then swim, accompanied by parents, across Baffin Bay to West Greenland. The spring of 1978 was exceptionally late, the summer exceptionally cold. Vast fields of ice remained in Lancaster Sound and tens of thousands of murre chicks died.

If the unseasonal presence of ice was fatal to the murres in 1978, the absence of ice in the Gulf of St. Lawrence or off Newfoundland and Labrador during two vital weeks in March can doom an entire cohort of harp seal pups. Female harp seals haul out upon ice to bear their young. If there is no ice, the females can delay birth for a few days while they desperately search for it, as they did in 1969 when there was virtually no ice in the Gulf of St. Lawrence. Many pups were born at sea and died immediately. Others were born on thin, slush-covered shore ice. When a storm broke even this miserable cradle, nearly all the pups perished. In 1981, there was again little ice in the Gulf of St. Lawrence, and pup mortality was extremely high.

The worst famine in the hunger-haunted oral history of the Netsilingmiut occurred in the year ''when winter did not come'' (probably in the mid-nineteenth century). In the normal rhythm of their lives, the people moved in early winter onto the ice, lived in igloos, and the men hunted seals at their breathing holes. But in that tragic year, the weather remained mild and ice did not form. Weeks passed, and months, food caches were depleted and the unfortunate people starved, and the living ate the dead. Finally, in late January, it became intensely cold, the sea froze over, and the remaining men could hunt seals again.

Such vagaries of weather were once fatal to arctic man and are still fatal to northern animals. Local populations of some animals may vanish. Others may be decimated and require years, even decades, to reach their former abundance. The vast reaches of the Arctic, however, are not affected, because most natural disasters tend to be sporadic and local, their impact limited in space and time.

The impact of man, though, was felt in the entire North. By an ironic twist of fate, the very materials that enabled arctic animals to withstand the severe cold of their world, their fat and fur and feathers, led to their extensive destruction throughout the Arctic, for these were all products coveted by southern man.

Arctic fox in thick white winter fur. This fur offers such perfect protection, arctic foxes are comfortably warm even at −40° F.

The Fatal Harvest

A when ye entered, ye defiled my land, and made mine heritage an abomination. JEREMIAH 2:7

RCTIC man hunted in order to live. Agriculture and animal husbandry freed southern man from this primal imperative. The hunt of arctic animals, which he pursued so resolutely and so ruthlessly for nearly a thousand years, and which reached its apogee in the late nineteenth century, provided him primarily with luxury goods he greatly desired.

In 890, the Norse chieftain Othere who "dwelt furthest North of any other Norman" (in Norway) visited King Alfred the Great of England and told him of his travels and his wealth. He had sailed as far north and east as Russia's White Sea to hunt "horse-whales" (walruses). Their ivory tusks were greatly in demand then (as they are now) for everything from crucifixes to knife handles, and walrus skin was "very good to make cables for shippes" (walrus thong, according to anthropologist Froelich Rainey, was "the strongest line known before the invention of the steel cable").

Although Othere was a great lord in his land, he had "but 20 kine and 20 swine." What made him wealthy was the tribute he received from subjugated Lapps: bird down by the bushel, seal skins, "whale bone" (baleen), and, above all, furs.

Furs, then as now, meant warmth and wealth and very gratifyingly combined comfort with status. Homespun clothing was not warm. Homes in winter were uncomfortable: the castles of the great were notoriously draughty; the merchants' houses chilly; the peasants' wattle and daub hovels miserably cold. Those who could afford it, the rich burghers of France or the boyars of Russia, dressed in furs. They were warm and helped to ease the prevalent rheumatism of middle age in the Middle Ages.

Furs also denoted rank. Only great nobles were allowed to wear sable, and ermine is still a badge of nobility. Chaucer described the rich merchant who wears a magnificent "bever hat," while the poor had to be content with "mittens of . . . curres skin." The best furs came from north and east, from Scandinavia, Poland, and Russia. Demand was great and prices high, for Europe's fur-bearing animals, first victims of this greed for northern furs, were becoming increasingly scarce. The discovery of Canada was most propitious.

Although Voltaire dismissed Canada as nothing but "a few acres of snow" not worth fighting over, Europe's rulers and merchants knew better: from this vast land fortunes in furs flowed for centuries to the Old World. When the Comte de Frontenac ruled as governor of New France in the late seventeenth

---

Walruses use their ivory tusks not to dig up food, but to fight and impress each other. Bulls with the longest tusks are, as a rule, at the top of the walrus hierarchy.

ABOVE: *Once hunted to near-extinction, the Pacific walrus, with careful management, has greatly increased in recent decades.*

RIGHT: *A newborn harp seal pup in pure white, down-soft natal fur. The immense appeal of the "cute" pups was a major factor in ending the harp seal hunt.*

century and *coureurs de bois* roamed far through the infinite forests to obtain furs from the Indians, 500,000 beaver pelts were sent annually to France.

The English, barred by the French from a southern approach, decided to tap this immense fur wealth from the north. On May 2, 1670, King Charles II signed the charter of "The Governor and Company of Adventurers of England Trading Into Hudson's Bay," better known as the still-flourishing Hudson's Bay Company, making them "true and absolute Lordes and Proprietors" of roughly 1,500,000 square miles, a fur empire nearly thirty times larger than England. From its posts on Hudson Bay, hundreds of shiploads of wilderness wealth went to England, the pelts of beavers, otters, martens, lynx, wolverines, bears, wolves, even moose hides and deer skins.

They pursued not only fur-bearing animals. Churchill, on the west coast of Hudson Bay, began its existence in 1688 as a combined whaling-trading station when the Hudson's Bay Company sent there "a Good Shipp [with] a competent cargo for Trade & Materialls, for White Whale ffishings." The next year, twenty-eight casks of white whale oil were shipped to England. On June 17 of that year (1689), the Hudson's Bay Company trader-explorer Henry Kelsey spotted northwest of Churchill two "ill shapen beast Their Body being bigger than an ox." The muskox had been discovered. At that time it had no commercial value. But once the bison had been nearly exterminated, muskox robes were suddenly in great demand. They made superb sled blankets. Between 1862 and 1916 the Hudson's Bay Company alone bought more than 15,000 muskox pelts, and whalers and other traders bought thousands more.

At first the Hudson's Bay Company was content to occupy posts at strategic river mouths on Hudson Bay and James Bay and let the Indians come down the rivers to them, their birchbark canoes laden with furs. After 1763, the energetic, Montreal-based North West Company spoiled this by the simple expedient of setting up posts in the hinterland that siphoned off the furs the Indians had hitherto brought down to the bay. The Hudson's Bay Company responded by establishing posts even farther inland, and for the next half century (they merged in 1821) the two great rivals leapfrogged west and north to reap the golden fur harvest of a virgin land. On July 14, 1789, the day that in Europe an angry Parisian mob stormed the Bastille, Alexander Mackenzie of the North West Company, after descending the great river named after him, reached the arctic sea. Four years later, he crossed the Rocky Mountains and reached the Pacific. A continent had been spanned in the quest for furs.

The exploitation of Canada's fur wealth inland and its fish wealth off the east coast produced some incidental victims. Funk Island, in the icy Labrador Current off Newfoundland, where the great auk nested in such numbers that, as Jacques Cartier wrote in 1535, "all the ships of France might load a cargo of them without one perceiving that any had been removed," became the larder of the European fishing fleets. By 1800 the great auk on Funk Island had been exterminated.

A polar bear mother and her yearling cubs near the edge of Hudson Bay. The cubs stay with their powerful, protective mother for two to three years.

The far-south walrus colonies on the Magdalen Islands in the Gulf of St. Lawrence and on Sable Island off Nova Scotia were wiped out. Walrus hides, noted the historian Richard Hakluyt in 1600, were "excellent good" to make shields "against the arrowes of the Savages" and London's "combe & knife-makers" paid well for their ivory tusks. The last walrus on the Magdalen Islands was killed in 1799. The white whales in the St. Lawrence River were avidly pursued. They were good to eat, and their hides, wrote the Jesuit historian Charlevoix in 1720, were ideal "for covering coaches." Eskimo curlews were killed in millions. They are now nearly extinct. And the Inuit of southern Labrador who interfered with fishing were exterminated.

Because gyrfalcon competed with man for ptarmigan, governors of the Hudson's Bay Company at Churchill, reported the explorer Samuel Hearne, used to give the Indians "a reward of a quart of brandy for each of their heads." Increasing literacy in Europe created a growing demand for quills as pens. In 1828, the Hudson's Bay Company sold 347,298 goose, swan, and eagle quills. Swans' down became popular for powder puffs and cozy muffs. Between 1823 and 1880, the Hudson's Bay Company sold 108,000 swan skins in London. By 1916, the magnificent trumpeter swan, largest of all waterfowl, was on the edge of extinction. Only about 100 were left.

The fur empires of the West, though vast and wealthy, were surpassed on both counts by those of the East. In 1581, Cossack troops broached the "Iron Gate" of the Ural Mountains and swept across the crumbling Khanates of Siberia. Sixty-seven years later they reached the Pacific Ocean. The wealth of Siberia lay at Russia's feet, and she did not delay in claiming her tithe. In 1590, eleven years after the conquest of Siberia began, Moscow demanded from its new dominion an annual *yassak* (tribute) of 200,000 sable skins, 10,000 fox skins, and 500,000 squirrel skins. By the middle of the seventeenth century, Siberian furs–of sables and ermines, of foxes, squirrels, bears, wolves, martens, mink and lynx, even of tigers and leopards from the Ussuri region–accounted for one-third of Russia's total income.

In 1741, Russia continued her eastward expansion. The Danish-born explorer Vitus Bering, in the service of the tsar, discovered the Alaskan mainland and the Aleutian Islands. On the return journey his ship broke up on the bleak Commander Islands off Siberia and there Bering died. Thousands of arctic foxes lived on these islands where no man had ever been. Totally without fear, they were easy to kill. A flightless cormorant existed on the Commander Islands and nowhere else in the world. Three-ton sea cows, giant northern relatives of the tropical manatees and dugongs, placidly munched seaweeds in the shallows. Sea otters lolled among the kelp beds, tame and trusting. And in spring, the fur seal came in tens of thousands to their island rookeries to breed.

The survivors of the expedition returned to Siberia in a boat made from the remnants of their ship, with a cargo of sea otter pelts. On the fur-hungry

RIGHT ABOVE: *The muskox's defensive circle was a perfect stratagem against their main natural enemy, the wolf. When confronted by men armed with rifles it was suicidal, and entire herds were gunned down.*

RIGHT BELOW: *Muskox bull in winter. Long guard hairs and dense underwool make adult muskoxen nearly impervious to arctic cold.*

Chinese luxury market, each of these pelts was worth more than $100, then the equivalent of a Russian worker's annual salary. By 1823, the value of one sea otter pelt had soared to several hundred dollars.

Siberia's *promyshlenniki*, the fur hunters, responded instantly to this promise of wealth. They were a breed of northern conquistador, ruthless and rowdy, brutal and brave, and furs were their gold. Using any conveyance that could reasonably be expected to float, they rushed to the new land. Potav Zaikov set out with seventy men in 1774 and returned seven years later with fifty-eight men and the pelts of 4,376 sea otters, 3,449 foxes, 92 otters, a wolverine, 3 wolves, 18 mink, 1,725 fur seals and 9¼ poods (334 pounds) of walrus ivory, worth a total of 300,416 rubles. Estimates of the total number of sea otters killed between 1741 and 1911 range as high as one million. When, in 1911, the sea otter finally received total protection, it was, according to biologist Karl Kenyon, "commercially extinct and nearly extinct as a species."

In 1786, Gerassim Pribilof discovered the Pribilof Islands in the Bering Sea with their immense fur seal colonies and between 1789 and 1867, two and a half million fur seals were killed. Millions more were killed after the United States bought Alaska from Russia in 1867 for $7,200,000. In less than twenty years, net revenue to the United States treasury from the sale of fur seal pelts alone paid for this land, twice the size of Texas.

As in Canada, the frenetic, immensely profitable hunt for northern furs claimed incidental victims. The sea cow of the Commander Islands, discovered in 1741, was extinct by 1768. The flightless spectacled cormorant of the same island group soon shared its fate. And although Catherine the Great of Russia enjoined the *promyshlenniki* "not to molest and not to cheat their new brothers, the inhabitants of those islands [the Aleutians]," these violent men paid little heed to such humane edicts. In a few decades, the once-proud Aleuts had been reduced from an estimated 20,000 to a broken, subservient remnant of barely 2,000.

Fortunes in furs came from the lands of the North. Even greater wealth came from its seas. Arctic whaling on a vast scale began after Henry Hudson, foiled by ice near Spitsbergen in his attempt to sail to China across the top of the world, returned to London in 1607 with rapturous reports of the whale wealth of this region.

These were bowhead or Greenland right whales, "right" because they were the whalers' ideal whale: slow, timid, colossal, and so fat they floated when killed. Thirty tons of blubber could be peeled off a large whale and rendered into more than 3,000 gallons of oil for the lamps of Europe and America. In its cavernous mouth it carried a ton of baleen, the "whalebone" greatly in demand in the days before plastics for everything from umbrella ribs to women's stays and once worth $6 a pound. So valuable was the bowhead that a single whale would repay the cost of a year-long voyage and yield a profit besides.

Young Steller sea lions play-fight on a wave-splashed rock.

TOP: *The kiss of recognition. After returning from the sea, a harbour seal mother sniffs her pup, recognizing it by its smell.*

ABOVE: *Hooded seal "family": the female with her pup and, behind them, a male with inflated "hood," the skin crest to which this species owes its name.*

A female harp seal surfaces in a breathing hole to look for her pup on the ice. Harp seals breed on the southern pack ice in spring and migrate to the far northern seas in summer.

Hudson's glowing report lured ships of many nations to the North: Danish, Dutch, German, English, French. The Dutch were the most successful, and their profits, for that age, were astronomical. Between 1675 and 1721 they employed 5,886 ships in the Spitsbergen "fishery." They took 32,907 whales, with a gross value of $82,267,000. In addition, the whalers killed thousands of polar bears, tens of thousands of seals, and wiped out the once-vast walrus colonies of the Spitsbergen archipelago.

After this region had been "cleaned out," as the whalers so aptly put it, they moved west to the Greenland coast (where they augmented profits by killing harp and hooded seals) and then through Davis Strait into Baffin Bay and immensely rich new whaling grounds. The mid-nineteenth century was the golden era of American whaling. In 1846, 735 American vessels (and 230 British and European ones) were hunting whales around the globe. In 1854, the best year, they returned with 10,074,866 gallons of bowhead oil and 3,445,200 pounds of baleen.

Hundreds of ships with thousands of men pursued the rapidly dwindling bowhead whales of the eastern Arctic. Whenever possible, they obtained "local food." One ship stopped at a murre colony on the west Greenland coast and in a few hours the crew collected "a thousand dozen eggs." Near the turn of this century, when the bowhead had been nearly exterminated in the eastern Arctic, the whalers, to make voyages pay, took from the North all that could be turned into money. They killed seals, polar bears, walruses, and thousands of white whales. They obtained fox fur and muskox robes from the natives. In 1904, the *Active* sailed for home with a cargo consisting of two bowhead whales, nineteen white whales, thirty-eight walruses, fifty-two seals, thirty-two bear skins, one hundred fifty-eight fox skins, thirty muskox hides, and one gyrfalcon.

The whalers' sweep through the western Arctic did not last as long but was even more intense and equally devastating. Between 1835 and 1915, more than 5,000 ships with 150,000 men took part in this hunt. They scoured the northern Pacific, the Bering Sea, the Sea of Okhotsk, and the Chukchi Sea and finally, in 1890, invaded the Beaufort Sea, last refuge of the hard-pressed whales. They reduced the bowhead from tens of thousands to a pathetic remnant of less than 3,000 in the entire Arctic. Of an estimated 300,000 Pacific walruses, only 45,000 survived the slaughter. The whalers paid Inuit to provision the whaling fleet with fresh meat, at the rate of 10,000 pounds per ship. Caribou, moose, muskoxen, even Dall sheep were indiscriminately slaughtered. "The district west of the Mackenzie [River] is now practically free of caribou," the Royal North West Mounted Police reported in 1914.

The whalers took from the Arctic much of its wildlife wealth, and they brought to the Arctic diseases to which the hitherto isolated Inuit had no immunity. Between 1850 and 1885, the Inuit population of the arctic Alaskan

Polar bears at nightfall in early winter upon the ice of Hudson Bay.

coast declined by 50 per cent. Within two decades of the whalers' arrival, 90 per cent of the Mackenzie Eskimos were dead. In the fall of 1902, whalers brought illness (possibly typhoid) to the Sadlermiut of Southampton Island in northern Hudson Bay. That winter the entire population died. The whalers' arctic reign lasted three centuries. They left a land and a sea despoiled and a people racked by disease.

Most of the great commercial hunts for arctic animals have long since ceased. There simply were not enough animals left to make them profitable. Only seal hunts, now regulated by quotas, persist: for hooded seals on the ice off Labrador and Greenland, for harp seals, of which, during the past 200 years, more than seventy million have been killed, and for fur seals at their rookeries in the north Pacific.

Though upset by the bad "image" given to these hunts by conservation groups–in the 1970s the United States Department of Commerce awarded a $49,910 contract to the Batelle Institute of Ohio to devise a "more aesthetic method" of killing fur seals–proponents of the hunt insist the killings are essential not only to provide badly needed income for Aleuts and poor Newfoundland fishermen but because the seals compete with man for fish. But finally the conservationists won by throttling the market. Without an outlet for furs, most sealers ceased the hunt. Only 826 harp seals were killed in the Gulf of St. Lawrence in the spring of 1985.

Persistent overfishing by the great fishing fleets of the Soviet Union, Japan, and Korea in the Bering Sea is probably a major cause of a recent decline in fur seal survival. In addition to this indirect threat to the seals, pieces of long-lasting nylon netting, cast adrift by fishermen, strangle many fur seals and sea lions. Birds, too, are affected by the fishing that seems to reach farther and farther north each year as southern fish stocks become depleted. According to a report by the Canadian Wildlife Service, "between 1968 and 1973, from 500,000 to 750,000 murres were killed in Danish fish nets off Greenland each year."

Man, traditionally, has eliminated real or suspected competitors. In Alaska, for more than forty years, bounty was paid on bald eagles because they were accused of eating too many salmon. Between 1917 and 1952, more than 125,000 eagles were killed, and "in the remoter hamlets of Alaska, eagle claws were legal tender," reported the biologist Victor B. Scheffer. When caribou declined drastically across the Arctic in this century, it was admitted that overhunting by man was the main cause. Nevertheless, as much blame as possible was shifted onto the wolf, and tens of thousands of wolves were poisoned, trapped, or shot. A bounty on their heads and a high price for their skins encourages the killing. In the USSR wolves are seen as a menace to the valuable reindeer herds, and they have been machine-gunned from low-flying planes.

Two male polar bears in a friendly embrace. Young males often play-fight, training, perhaps, for the real fights with rivals during future mating seasons.

Our eternal yen for exotica claims many arctic victims. A lethal corollary to this craving is the rule that an animal's value increases with its rarity. The 1892 "Standard Catalogue of North American Bird Eggs," the oologists' bible, listed these prices for eggs: trumpeter swan (then nearly extinct) $4 each; heath hen (now extinct) $3 each; whooping crane $3 each. Narwhal tusks sold for $20 to $30 in the 1930s. Now a seven-foot tusk is worth at least $2,000, and an exceptionally long and beautiful tusk was recently sold at a London auction for $7,000. As a result, narwhal are now hunted primarily for their valuable tusks. Muskoxen are shot because tourists pay $500 and more for a "nice" set of horns. And the main incentive to walrus killing is the great value of their ivory tusks. They are carved into high-priced tourist trinkets or, as the biologist Karl Kenyon observed, are traded by the natives "over the bars in Nome [Alaska] for liquor-by-the-drink." Saudi princes covet the magnificent white gyrfalcons of the North. In 1981 a Canadian Inuit association asked the government of the Northwest Territories for permission to capture 400 of these birds, expecting from this a revenue of $12 million. In the 1970s, polar bear rugs became a status symbol in Japan, and the price of polar bear skins shot up to $3,000 each.

Polar bear gall bladders are prized as a panacea in the Far East. Their price is high and Korea is the main market. Fur seal genitalia, known in the trade as "seal sticks," are valued in the Orient as a reputedly potent aphrodisiac. Hong Kong merchants pay well for them. Sea lion whiskers are in demand. Nothing, it seems, surpasses them for cleaning opium pipes.

Infinitely more serious than such odd and incidental stimuli to increased hunting of certain arctic animals are three factors that will determine the future of arctic wildlife: widespread industrial exploration and exploitation of the North's mineral resources; the natives' insistence on hunting, preferably without limitations, as their birthright and as the *sine qua non* of their cultural survival; and a philosophy that views animals primarily in terms of their potential monetary value.

The natives of the North argue with great cogency that it is their land, that whites have been largely responsible for the initial decimation of arctic wildlife, and that hunting is an integral and essential part of their way of life. While justified, these arguments skirt the fact that northern animal resources are limited, that human populations in the North are rapidly increasing, and that snowmobiles, high-powered rifles, speedboats, and even aircraft now make hunting so efficient that without some controls the most avidly hunted species, such as the caribou, are bound to decline. It is probable that only a deep involvement of the native people in the process of protection and perpetuation of arctic wildlife will halt and perhaps even reverse this fatal trend.

The invasion of the Arctic by industry poses a plethora of problems, present and potential. A single major oil spill in northern Baffin Bay, David Nettleship of the Canadian Wildlife Service has warned, could destroy "as much as three-quarters of the total murre population of Lancaster Sound and Jones Sound,'' two of the richest seabird areas in Canada. It would presumably also endanger this area's walruses, white whales, the few remaining bowhead whales, the seals, and the narwhal. This region, which, because of its animal wealth, is often called an ''arctic oasis,'' could become a barren desert.

In a moment of candour, which he probably later regretted since he was quoted in the press, the president of a northern exploration company said: ''We're not really all that interested in the scenery and the animals. What we want to do is to make some money out of it.'' Though somewhat blunt, this statement expressed accurately the priorities of industry whose quintessential task and goal in the industrial state is to be profitable and which consequently views conservation as an impediment to profitability.

So far industry's impact upon the Arctic has been minor and restricted: a mine site here, a road there, a lake killed by tailings, minute scabs on the infinite face of the North. But each year this blight spreads a little, and there is a distinct danger, as the biologist Ian McTaggart Cowan has warned, that the fragile northern environment will be destroyed in ''incidental increments.'' It will be the challenge of the next decades to develop the North without destroying this last great wilderness on earth. ''We, speaking for mankind collectively,'' wrote the famous economist and author Stephen Leacock in 1936, ''have for the present at least made a mess of the rest of the world. For the North let us make it different . . . let us see to it that in the new trust of the future of the North, we make fewer errors than in the old.''

Taiga — Treeline — Tundra

THE treeline is the border between two of the world's great biomes: the taiga, the austere, primarily coniferous boreal forest, and the immensity of the treeless tundra to the north of it. Temperature determines the treeline's course; it coincides roughly with the 50° F. isotherm for the warmest month of the year. But taiga and tundra meld and merge in erratic, interlinking fashion as local conditions of soil, moisture, humidity, elevation, precipitation, and permafrost favour or foil the growth of trees. Islands of tundra occur deep within the boreal forest, and oases of trees cluster in sheltered tundra valleys far north of the treeline.

In Scandinavia, the most common trees of the North are spruces and pines. But birches dominate the treeline region, and Lapland is glorious in fall, its forest shimmering in silver and gold, the ground covered with a thick, patina-green carpet of reindeer moss.

Throughout Siberia, the light and lofty larches rule the northern forests; they cover slightly more than a million square miles. A gentle green in spring and summer, they stand stark and skeletal in winter because, unlike other conifers, larches shed their needles in late fall. Incredibly hardy, larches can endure temperatures of −90° F. They reach their northern limit in the valley of the Khatanga River, near the base of the icy, wind-swept Taymyr Peninsula, and a few isolated copses exist even farther north, at the latitude of northern Baffin Island. But there their growth is exceedingly slow, on the average less than half an inch a year.

Boreal forest covers about half of Canada and Alaska; its dominant conifers are the white and black spruces, balsam fir, and tamarack. This brooding, spired forest becomes increasingly open and parklike in the North. Near the treeline, mop-headed black spruces predominate. Shallow-rooted, they can survive even where permafrost, the concrete-hard layer of eternally frozen soil, and naleds, bulges of nearly pure ice, are only a few feet beneath the surface. But poorly anchored and pushed by buckling sub-soil ice, the trees lean crazily, and many lie toppled. Botanists call it "the drunken forest."

At the treeline only black spruces survive. German botanists with a flair for the dramatic call the treeline a *Kampfzone*, a battle zone, and the battle is for survival. Flayed and frayed by severe winter storms, their buds seared by the ice-hard spicules of wind-driven snow, the spindly spruces look like gale-torn weathervanes. Some of these trees, only wrist-thick and eight feet tall, are centuries old. Here at the border of possible life, the growth rate of their trunks

A fishing Kodiak bear in Alaska stares at the salmon that got away.

is only one-twentieth of an inch per year–in good years. In bad years they do not grow at all.

Taiga and tundra are totally different realms. The forest encloses, shelters, protects. Wind intensity decreases by 90 to 95 per cent only fifty yards from the forest edge. On warm summer days the calm forest air is fragrant with the scent of resin and the earthy smells of moist mosses, muskeg, and mouldering duff. The tundra is open and spacious, awesome in its immensity, a land as vast and lonely as the sea. In winter, snow lies deep, soft, and fluffy within the sheltered forest belt, and Indians travel on snowshoes and with flat-bottomed toboggans. On the tundra, winter storms sculpt and compact the snow until it is so hard the Inuit's sharp-edged sled runners leave a barely perceptible track, and snow can be cut into blocks and made into an igloo.

Once, the treeline was the division between human communities. The Indians, essentially, were a people of the forest, while the tundra and the northern coasts were home to the Inuit. The treeline is also a border for many animal species. The snowshoe hare and its arch-enemy, the lynx, the red squirrel and its twin foes the marten and fisher (and sable in Siberia), the crossbills, siskins, thrushes, woodpeckers, chickadees, kinglets, and grouse venture to the forest edge but rarely beyond. To other species, ptarmigan, muskox, arctic fox, and snowy owl, the tundra is home. And a few, like the caribou, commute, spending summer on the tundra and winter within the forest belt.

The division of species can be quite abrupt. Just south of the treeline, 10,000 species of insects exist in the boreal forest; just north of it, on the tundra, only 500. Unfortunately, the two main insect scourges of the North, blackflies and mosquitoes, which torment man and beast often to near-madness and occasionally to death, fill with their legions both tundra and taiga.

Blackflies are tiny but terrible. Only the female bites, tearing out minute chunks of flesh with saw-edged mandibles. Into the wound she injects an anticoagulant to keep the blood, which she requires to nourish her eggs, flowing. This saliva is toxic and bites bleed freely, itch terribly, swell, smart, and often become infected. When the explorer George Back travelled through the region northeast of Canada's Great Slave Lake in the summer of 1833, blackflies "rose in clouds . . . darkening the air. . . . Our faces streamed with blood . . . there was a burning and irritating pain, followed by immediate inflammation and producing giddiness, which almost drove us mad . . . even the Indians threw themselves on their faces, and moaned with pain and agony." Blackflies are also incredibly prolific. Researchers in northern Quebec counted blackfly eggs on a fifteen-foot stretch of rock beside a waterfall and estimated their number at sixteen billion.

Equally numerous, nearly as painful, but easier to ward off, are the North's mosquitoes. The clouds of mosquitoes that swirl like glittering smoke in the rays of the evening sun consist of males and they do not sting. Females can live

An arctic fox in brownish summer fur. The winter pelage is pure white. Along coasts, arctic foxes feed mainly on seabirds and their eggs. Inland, lemmings are their principal prey.

on plant juices but blood enhances their egg production. A single female sucks in only two to eight milligrams of blood per bite. But a naked man on the tundra, according to British mosquito expert J.G. Gillett, "could receive more than 9,000 bites per minute, which would result in the loss of half his blood in less than two hours." After nearly dying of cold and scurvy during a winter (1631-32) spent in the bay that now bears his name, the explorer Thomas James and his crew, when summer came, were beset by "such an infinite Number of blood-thirsty Muskitoes, that we were more tormented with them, than ever we were with the cold Weather."

Only wind and cold bring relief from this mid-summer plague. On windy days, mosquitoes and blackflies hide near the ground. At 50° F. they become sluggish; at 45° F. they are torpid and cannot fly. At all other times during June, July, and well into August, they fly and buzz and bite. They can kill small mammals, brooding birds, and fledglings, and sometimes they harass birds so badly that the birds abandon their nests. They torment caribou, who seek out wind-swept ridges and patches of snow to escape the clouds of mosquitoes. Instead of grazing, the desperate animals run, and as biologist George Calef has pointed out, their "efforts to escape the insect pestilence cost them precious energy at a time [after spring migration and calving] when their reserves are already low." In addition, "each animal may lose up to a quart of blood a week during the peak of the insect season."

When one is covered by mosquitoes and bitten to distraction, it is hard to see any good in them. But they do form an important strand in the food webs of the North. They and their larvae are eaten by predatory insects; they feed a host of fishes and a myriad of birds that come to the North in summer. The appropriateness of all things is more evident when one awakes on a cool tundra morning in a mosquito-speckled tent and watches snow buntings dance on the canvas as they pick off the numbed insects in droves and carry them to their young.

Except in the northern portion of the boreal forest where trees are often widely spaced, the taiga is a gloomy place. The wind sighs through the treetops, but beneath the blue-green canopy the air is calm and cool; in this deeply shaded realm few plants can prosper. Mosses, lichens, some winter-greens, and horsetails cover the forest floor, and young spruces struggle up-wards towards the life-giving light. Flocks of chickadees and kinglets flit and flutter high in the trees, busily picking insects, their larvae, and their eggs. Crossbills use their strangely shaped mandibles to open conifer cones. They pry the scales apart and extract the seeds with scoop-shaped tongues. Crossbills inhabit the boreal forests of North America and Eurasia, and they are revered in European folklore. According to the ancient tales, when Christ was crucified on Golgotha, crossbills tried to pull out the nails. They twisted their bills forever and their plumage remains red-stained with the Savior's blood.

Spruce grouse, superbly camouflaged in black and brown and beige, fly up with a clutter of short, rounded wings, then sit on low branches and cluck in wonder. "Fool hens" they are called in the North for they are tame and trusting, a fatal trait when they meet man. But since they eat mainly spruce needles, buds, and shoots, their dark meat has a peculiar, bitter resinous flavour. They are one of the few bird species to winter in the taiga. In the evening they dive deep into the powdery snow and spend the icy winter nights beneath this insulating blanket.

The most conspicuous and noisy inhabitants of the northern forest are red squirrels. They chitter and chirp and scold furiously when disturbed. These squirrels are compulsive hoarders. In peak years, some squirrels lay by enough cones to last them through two winters. Much of this wealth they cache within their middens, three-foot-high piles of bracts and cone cores that can be thirty feet across and are often the accumulated refuse heap of many squirrel generations. The squirrel sits upon a branch above its midden busily shucking cones. As it eats the nutritious seeds, a steady rain of discarded scales descends upon the ever-growing pile.

Red squirrels are extremely possessive about their middens. When you see squirrels racing through the treetops in what looks like a merry chase, it is most probably an infuriated midden owner running a suspected thief off his property. The midden is both lair and larder. The squirrel stocks it with food in summer and sleeps in it in winter. But many do not survive to eat their ample hoards, for red squirrels are avidly hunted by marten, mink, fisher, lynx, fox, owls, and hawks.

Brown in summer, white in winter, and mottled in spring and fall, the snowshoe hare of the northern forests changes colour to blend with the seasons. Thanks to its outsize, splay-toed, thickly furred hind feet, to which it owes its name, this little hare (it only weighs three to four pounds) can run with ease on the fluffy-soft snow of the forest. So, to the hare's misfortune, can its main enemy, the lynx, which also has large, thickly furred paws.

Snowshoe hares prefer brush country, especially areas cleared by forest fires that are rich in such successional trees as poplar, aspen, willow, and alder. In summer they eat buds, shoots, grasses, and leaves; in winter primarily twigs and bark. Stunningly fecund (females bear as many as ten young per litter), snowshoe hares can increase a hundredfold in just five years, and their populations soar and crash in ten-year cycles. It was long believed that these catastrophic die-offs were the result of overbrowsing, overcrowding, and disease. Now Alaskan scientists have discovered another reason for the hares' abrupt demise. When they become too numerous, the very plants upon which they depend for food fight back by producing toxins that act as hare repellents. Deprived of food, the animals starve. The land that swarmed with snowshoe hares is suddenly empty. Famished lynxes prowl far and wide, an easy catch for trappers, until they, too, succumb and die. Then the plants cease to be toxic; the remaining hares have ample food and multiply again, and the ten-year hare-lynx cycle begins anew.

While hares and lynxes are, to some extent, victims of food specialization, bears avoid this by being omnivorous. It is surprising to see Alaska's giant brown bears, the largest land carnivores on earth with the possible exception of the polar bear, browse placidly on northern meadows munching sedges and grasses. They also eat shoots, roots, and berries, and carrion when they find it. Large males weigh three-quarters of a ton, and some, when they rise on their fourteen-inch-long hind paws, stand ten feet tall.

By inclination, brown bears are loners. For much of the year they roam the wilds, solitary and sour-tempered, often on well-worn trails that should be avoided by hikers. Brown bears rarely attack humans, but when they do they are terrifyingly fast and their strength is awesome. The biologist George Schaller reported that one brown bear hauled with ease the 1,000-pound carcass of a drowned moose out of a river and onto the bank.

In mid-summer, when salmon ascend Alaska's rivers, brown bears congregate at favourite spots to catch the migrating fish. About eighty bears from an enormous region converge on McNeil River Falls on the Alaska Peninsula, the most famous of their fishing places. At the falls, these grouchy individualists must somehow get along, for the six-week salmon feast is extremely important to the bears. They round out visibly on this rich diet, gaining as much as seven pounds each day, and accumulate the vital fat layer that will sustain them during hibernation.

To avoid fights and strife that consume precious energy and time, the bears fishing near the falls adhere to a strict order of precedence. At the head of this hierarchy are the strongest males. They occupy the best fishing places and return to exactly the same spots year after year until at last, old and ailing, they are displaced by stronger rivals. Inferior fishing places are held by bears of lower social standing: females with cubs, smaller males and females, sibling groups, and, on the lowest rung, lone immatures. Since each bear knows his place, fights are relatively rare and all can concentrate on the serious business of catching fish.

Whenever a salmon tries to slip past a bear, he pounces upon it with a speed and agility incredible in an animal so big and bulky, pins it down with long-clawed paws, and grabs it nearly simultaneously with his mouth. On good days a big bear catches twenty or more salmon, weighing an average of six pounds each (one biologist saw a bear catch sixty-nine salmon in one day). At first the bear eats all the fish, then just the females, and finally, sated, takes only roe, leaving the rest to gulls, ravens, and grateful inferiors. By mid-August, the fishing season is over, and the bears disperse to harvest the ripening berries. At the onset of winter, swathed in fat, they seek out sheltered lairs and hibernate, living off their fat reserves during the food-poor winter months.

Its wings held high, an arctic tern settles upon its nest. Some of these dainty birds fly more than a million miles in their life.

The Sleeping Land

IN winter the tundra seems lifeless, a silent, solemn land, vast and white and void. *Nuna hiniqpoq*, say the Inuit, the land sleeps. In former times, in early winter, many Inuit groups left this land that had no food to offer and settled in small igloo communities upon the sea ice where the men hunted seals at their breathing holes. Armed with harpoons, they stood like statues for many hours, sometimes for days, and killed the surfacing seals with lightning strokes of their weapons.

Far on the ice, polar bears hunt seals with equal patience and persistence. In the vivid imagery of their poems, the Inuit call the polar bear *pihoqahiaq*, the ever-wandering one. Once, it was thought that polar bears roamed the top of the world from continent to continent, eternal nomads of the North. Some, in fact, do turn up in the remotest places. In May 1926, the explorers Lincoln Ellsworth and Roald Amundsen crossed the frozen Arctic Ocean in the airship *Norge* from Spitsbergen to Alaska. Near the ''Ice Pole'' (86° north latitude, 157° west longitude), ''the most inaccessible spot in the Arctic regions,'' they saw ''one lone polar bear track.'' Several expeditions in recent years have seen polar bear tracks near the North Pole, and I saw them at the base of Newtontoppen, the highest mountain of the Spitsbergen archipelago, amid an extensive ice cap. But the recapture of tagged bears and recent genetic and morphological studies indicate that most polar bears belong to geographically discrete populations.

While most bears pad silently across the winter ice in search of seals, guided to their breathing holes by an extremely acute sense of smell, pregnant females head inland. In late November they seek out favourite denning areas, perhaps in the very region where they were born, and excavate oval lairs in deep snow-drifts in the lee of ridges or river banks.

The cubs–usually only one cub if it is the female's first birth; two, as a rule, thereafter, and occasionally three–are born in December or early January and seem singularly ill-prepared for their mid-winter arrival. They are only the size of rats, weigh barely one and a half pounds, and are blind, deaf, and nearly naked. Since snow is an excellent insulator, the temperature inside the den may be forty degrees warmer than the temperature outside, and as in Inuit igloos, the entrance tunnel slopes upwards so warm air will not escape from the den. The tiny cubs snuggle into their mother's deep-pile fur and suckle her fat-rich milk. She reclines on her back and cradles the cubs upon her chest with her massive furry paws. The cubs grow quickly. When they emerge from the

Though normally slow and ponderous, muskoxen can move with amazing speed and agility when fleeing or charging.

darkness of their den into the dazzling glitter of the Arctic in March, they are densely furred and weigh a chubby twenty pounds.

In early summer, when ringed seals and bearded seals bask upon the ice, polar bears stalk them with infinite patience and finesse, synchronizing their cautious approach with the sleep-wake rhythm of the seals. When the seal sleeps, the bear crawls forward with utmost caution. When the seal awakes, the bear freezes, a faded-yellow lump upon the snow-streaked ice. Nothing moves, all seems safe. Reassured by its careful survey, the seal falls asleep again and the bear resumes his patient stalk. At twenty feet he pounces and grabs and kills the seal before it can slide off the ice into the sea.

The bear's strength is prodigious, and its claws are so long and sharp that it can kill a 700-pound bearded seal with one mighty swipe of its paw. In 1970 on Jones Sound, near Ellesmere Island, the scientist Milton Freeman reported that a 350-pound female bear killed a white whale trapped in a *savssat*, a hole of open water surrounded by a vast expanse of solid ice, yanked the nearly 2,000-pound carcass onto the ice, and dragged it for twenty feet. In 1854, a polar bear found the food cache of explorer Elisha K. Kane and tossed eighty-pound cans about "like footballs." Using claws and teeth, the bear had opened heavy metal containers "as with a cold chisel."

Seals are the polar bears' main and favourite food. But when the ice melts, the bears must come ashore. Then they eat anything available: grasses, sedges, seaweed, fish, carrion, berries. They raid eider colonies and eat the eggs. On Spitsbergen in the seventeenth century, whalers buried their dead on an island. Afterwards, a contemporary chronicle reported laconically, "the white bears find them and devour them."

When seal hunting upon the ice is good, polar bears can be choosy feeders. They eat only the high-calorie blubber and skin, and they do this, as Ian Stirling of the Canadian Wildlife Service has observed, "in a very exacting manner." He watched a bear in the high Arctic who, after killing a seal, was "carefully using its incisors like delicate clippers to remove only the fat from the carcass, leaving the meat." Successful bears leave a trail of bounty for younger, less experienced bears, and for the arctic foxes who often follow them all winter in hopes of leftovers.

Wrapped in long, dense, silky-soft fur, arctic foxes in winter appear rounded and bulky. But beneath this warm and voluminous coat is a puny body; on the average they weigh only about ten pounds. Arctic foxes are dimorphic: nature produces both a blue and a white edition. Blue foxes are a deep smoky blue both in summer and in winter. White foxes are pure white in winter, and a brindled tawn, tan, and grey in summer. If a white fox and a blue fox mate, their offspring may be all white, all blue, or of both colours.

For reasons not yet understood, the "coastal" foxes of such regions as West Greenland, Jan Mayen Island, the Pribilof Islands, and the Commander Islands

---

Common murres stand shoulder to shoulder on a breeding island. Heavily dependent upon capelin, murres are threatened when too many fish are taken by man.

are nearly always blue, while 99 per cent of Canada's "continental" foxes are of the white phase. The coastal foxes eat mainly birds, while lemmings are the mainstay of the inland foxes. For the blue foxes living near one of the North's great seabird colonies, such as the immense scree slopes of northwest Greenland speckled with millions of dovekies, summer is one glorious feast that ends abruptly when the birds leave in August. The provident foxes provide for the long, lean winter by storing food during summer's brief season of plenty. The Danish scientist Alwin Pedersen found a fox's cache that contained thirty-six dovekies, two young murres, four snow buntings, and a large number of dovekie eggs. "The frozen bodies of the birds were neatly arranged in a long row . . . and the eggs were heaped in a pile. . . . The whole store would have provided food for a fox for at least a month."

Unlike the migratory birds, lemmings remain in the North all year and are a steadier source of food. But their numbers wax and wane in four- to five-year cycles, and fox populations, and those of many other northern predators that rely heavily on lemmings, rise and fall in rhythm with their prey. Lemmings look like chubby, oval, thick-furred mice with plump rounded rumps, short legs, tiny ears, minuscule tails, and black shiny button eyes. Normally shy and wary, they turn into tiny furies when cornered; Scandinavians say of a very courageous man that he is "as brave as a lemming." Chittering with rage, a two-ounce lemming, when flight seems futile, will attack a 1,000-pound polar bear, and a lemming pursued by men of Britain's Royal Navy on Melville Island in 1820, "set himself against a stone . . . and bit the serjeant's finger."

Inuit call the lemming *kilangmiutak*, the one that comes from the sky, and to Scandinavian peasants they were "sky mice," both groups accounting for the lemmings' incredible abundance in certain years with the assumption that they fell like rain from heaven. At such times, the tundra is aswarm with lemmings; they are as numerous as "the host of God," wrote Pontoppidan, an eighteenth-century Norwegian bishop.

The reason for this stunning abundance need not be sought in heaven for it is based on very earthy urges. Female lemmings are exuberantly prolific and extremely precocious. In March, when the awesome silence of the snow-clad tundra is broken by the hoarse screeching and yawping of courting arctic foxes, and high in the sky the raven woos his mate in wild and intricate flights, the lemmings are busy beneath the snow producing and raising the first of the year's litters. In a good year, a female can have six to seven litters, and the females of the first litter may mate thirty days after birth (one, in captivity, mated at the age of fourteen days) and produce young of their own after a twenty-day gestation period, with as many as eleven kits per litter. At this rate, one lemming pair can have thousands of descendants within a single year.

After one of their periodic population crashes, only one lemming may remain to every ten acres. The next year they are more numerous, extremely

---

Varying lemming in white winter fur. Immensely numerous in peak years of their cycle, lemmings are an essential food to many arctic predators and raptors.

common the third year, and in the fourth year, as the population soars towards another peak, they may number well over 100 per acre. In their legions, they improve the tundra soil. They aerate it with their burrows, as many as 4,000 per acre, and fertilize it with their droppings and their dead. But they can also denude the tundra, depriving other animals of food and shelter. A lemming eats daily about twice its weight in grasses, sedges, and other vegetation, or about 100 pounds in a year. "Fat, busy, agile mowing machines," the biologist Charles Elton called them.

Generally, though, peak lemming years are boon years for most tundra dwellers. Snowy owls have many owlets and manage to raise them all. A pair and its young eat about 100 lemmings a day. Arctic foxes have large litters. Jaegers, ravens and rough-legged hawks, wolves and ermines, gulls and eagles, all thrive on the lemming bounty. Even the meek, normally herbivorous caribou slays lemmings with its hoofs and eats them. Since lemmings provide plentiful food for predators, other prey species such as ptarmigan, sandpipers, snow buntings, ground squirrels, and arctic hares are spared and survive the season in much larger numbers than during years of lemming scarcity.

Lemmings are individualistic and quarrelsome. As populations grow, they become increasingly irritable and excitable. They squeak and fight furiously in their tunnels beneath the snow. Finally, hyperactive, their adrenal glands swollen due to the stress of overcrowding, the lemmings begin to disperse, not in serried ranks of suicidal marchers all heading to destruction in the sea, as these mass "migrations" are often described, but as a nervous, high-strung host of scattered, scurrying animals, all obsessed with the same frenzied urge to move. Some try to cross rivers and lakes, and although they are good swimmers, many drown or are eaten by trout. Others move into marginal food regions and starve. Their usually rampant libido declines, populations crash over wide areas, and for many tundra predators hard times lie ahead. Trappers take a record crop of hungry foxes. Snowy owls fly far to the South. Jaegers lay small clutches or may not breed at all. Wolves concentrate their hunting efforts on caribou, arctic hares, and ground squirrels.

Beset by a host of predators in summer, lemmings are safer in winter. The insulating blanket of snow shields them from the searing cold and from most of their enemies. Only foxes and ermines hunt them all year. A fox, alerted by the lemming's smell and telltale squeaking, digs rapidly into the snow, then suddenly jumps high and crashes down with stiff legs to break the snow and trap the rodent.

For the ermine, or short-tailed weasel, the hunt is easier. Lean and supple, it slides through the lemming tunnels. One hears the weasel's hiss, the lemming's squeak, a brief scuffle, and then silence. No lemming is a match for this lithe and lethal hunter that will tackle prey many times its size. In summer these nervy little animals are deep brown above and yellowish-white beneath.

In winter the weasel is pure white, only the tip of its tail is black. When an owl or hawk spots an ermine, white in a land all white, the main visual attraction is the black tip of its tail. Mislead by this attractive tail, the raptor strikes too far behind and the weasel can escape.

The impression that the tundra is lifeless in winter is partly due to the fact that whatever life there is is extremely hard to see. In this white and silent land, most animals are white and usually silent. The most startling exception is the raven, that "smartest and most cunning of all birds," as explorer-writer Peter Freuchen called it. Disdaining camouflage, it is glossy blue-black all year. It breeds even in the highest Arctic; it ranges widely in search of food, and it is incredibly hardy. In 1867, the whaler *Diana* lay locked in ice in Davis Strait. On January 31, when it was "terribly cold" (the temperature in the crew's quarters was $-22°$ F.) "a raven flew over the ship. . . . The bird's neck was encircled by a glittering ring of ice formed by the freezing of the moisture in its breath upon the feathers. As the bird approached us, this ring of ice sparkled in the sunlight like a diamond bracelet. It is wonderful how these hardy birds contrive to find a living in this awful place."

Like foxes, ravens follow polar bears in winter to eat whatever food they leave; they search leads and tidal cracks for dead fish and crustaceans; they follow Inuit hunters at a respectful distance; and nowadays they frequent the garbage dumps of northern towns, exploration camps, and military bases. Of all the northern birds, they are the first to court, mate, and nest. In March, while the North is still in winter's thrall, the ravens circle and soar in the cold-blue sky, and swoop and loop and barrel-roll in the ecstasy of courtship high above the frozen land.

Ptarmigan, the grouse of the North, are white in winter.

In summer the ptarmigan's plumage changes to rich brown.

The Rites of Spring

IT is an old saying that the Arctic has but two seasons: a long winter and a brief summer. In the high Arctic the plants' season of possible growth is forty days or less, and about eighty days on the arctic mainland. In 1971, when I lived with the Polar Inuit of northwest Greenland, the ice on Inglefield Bay broke up on July 23 and on August 10 we had our first snowstorm.

Spring in the North is tentative and tantalizing. Fierce April storms and days of bitter cold with temperatures as low as $-40°$ F. alternate with brief spells of pure enchantment. The air is cool and still and clear, the sun intense, the sky a deep robin's-egg blue shading to aquamarine near the horizon, the snow and ice aglitter. Evaporation haze shimmers and dances above dark ridges and rocks, and the snow slowly vanishes by ablation. "The Snow does not melt away here with the Sun or Rain . . . but is exhal'd by the Sun," noted the explorer Thomas James in the spring of 1632.

The gradual disappearance of the snow is vital to voles and lemmings, whose tunnels could be flooded by an abrupt spring thaw. Occasionally this happens, and the lemmings die in droves. Young drown in their nests, and wet, bedraggled adults flee to higher ground, an easy prey for predators. Even without flooding, many lemmings come to the surface of the snow in spring, especially in peak population years. Perhaps they simply can no longer tolerate close confinement beneath the snow with their swarming kith and kin. This spring emergence of lemmings is important to the ravens. The female does not leave the nest during the entire three-week incubation period, for her eggs would freeze instantly, and the male must hunt for himself and his mate.

In the northern forest belt where they have spent the winter widely dispersed in small groups, living primarily on caribou moss, the immensely abundant lichen that carpets the forest floor, the caribou in late March respond to an ancient call. As the days lengthen, they become increasingly restless. A few bands move north. Others follow. Trickles of animals meet and merge into a mighty stream of life pouring towards the tundra in that vast migration so vividly described in an Inuit poem recorded by Knud Rasmussen:

Glorious it is to see
The caribou flocking down from the forest
And beginning
Their wandering to the north.

A newborn caribou calf blends to perfection with the surrounding tundra vegetation.

Glorious it is to see
The great herds from the forest
Spreading out over plains of white,
Glorious to see.

To many natives of the North, the inland Inuit of Canada and Alaska, and to such Indian groups as the Chipewyans, and the Naskapi of Labrador, the caribou once meant life. They ate its meat and dressed in its fur. From its antlers and bones, they made tools, toys, and weapons. The Naskapi, said the explorer W.B. Cabot, who visited them in 1906, were "lords over their fine country, asking little favour, ever, save that the deer [caribou] may come in their time." When, in this century, caribou declined sharply in some regions, these people died or fled the land that for them had become truly barren. In 1947, the botanist Jacques Rousseau travelled through Labrador along the trail taken by starving Naskapi to reach the settlement of Fort Chimo. "*J'ai suivi sa lente migration vers la mort*" (I followed their slow migration towards death), he wrote. He saw only twenty caribou during the entire trip.

Once, the caribou may have numbered about three million. As recently as the 1940s, the Canadian biologist C.H.D. Clarke said that "it is to be hoped that there will never be so few caribou that it will be possible to count them." But soon scientists were counting and recounting them, and each successive census showed more starkly the havoc wrought by uncurbed and excessive hunting. "In just ten years, between 1965 and 1975," noted the biologist George Calef, "Alaska lost over half its caribou." The Kaminuriak herd of the Canadian North shrank from more than 150,000 in the 1950s to less than 40,000 now, and other once-vast herds declined nearly as rapidly, victims of unbridled and often wasteful hunting.

Biologists have named most of the major caribou herds after the geographical locations of their calving grounds. It is towards these remote areas where they were born that the pregnant cows trek so urgently in spring. Accompanied by yearling calves, the females march fifteen to twenty miles each day, probably guided across this vast and seemingly featureless land by clues and memories retained from past migrations. The bulls, who have no pressing date with destiny, follow far behind the herds of females at a more leisurely pace. The Oblate priest Raymond de Coccola watched in the 1930s the arrival of the herds near their final destination on the northern Canadian Barrens: "Through a delicate haze that rose from the snow-covered land under the warm rays of the sun, caribou were visible everywhere. They were flowing from every gully, pass, and ravine, fanning out into the broad valley, pausing to graze, and then slowly pushing on."

Once they reach the calving ground, the cows disperse. In the preceding fall, most females mated nearly at the same time, and now, in mid-June, most calves

---

The last rays of the setting sun catch a Baird's sandpiper busily probing mud exposed at low tide for worms and tiny crustaceans.

are born within a five-day span. This reproductive synchrony has great survival value. Following the birth of the calves, the herds can quickly regroup. Few laggards are left behind. To caribou there is safety in numbers. Apart from man, wolves are their main enemy. Although wolves, who are master strategists and hunt as a skilful team, can kill healthy caribou in relay chase or cunning ambush, they prefer, since it requires less effort and energy, to cull from a herd its most vulnerable and easily caught members: the old, the young, the weak, the sick. Laggards and strays are favoured victims.

But while the nearly simultaneous birth of the calves has great advantages, it also exposes them to one great danger. If during this vital five-day period a late-season blizzard blankets the land or, worse, freezing rain drenches the calves' fur, destroying its insulation, the just-born fawns may perish. In clement years, calves constitute as much as 25 per cent of a herd, and less than 5 per cent in disaster years.

At about the same time that the caribou calves are born, mosquitoes hatch in the tundra's myriad lakes and ponds. As if herded by this dreaded plague, the widely scattered caribou bunch again into large herds, and the vast throng moves across the land haloed in mosquitoes, the calves blatting, the females grunting and coughing, restless, harried, miserable, seeking breezy ridges to escape these pests, grazing or resting when wind or chill give them respite. The adults begin to shed their bleached, long-haired winter fur. They look moth-eaten, blotched, and haggard.

The mosquitoes have hardly begun to abate when the caribou are assailed by two insects they seem to fear even more: the warble fly and the botfly. Both look like slender, hirsute bees and their insistent buzzing drives the caribou mad. The warble fly lays its eggs on the fur of legs and abdomen. The larvae hatch, bore through the skin, burrow upwards, and lodge beneath the hide on rump and back. Some of the unfortunate animals are host to more than 1,000 of these thimble-sized maggots. The botfly is viviparous and deposits its larvae in the caribou's nostrils. They migrate from there to the throat, and victims wheeze and grunt and cough. Caribou try desperately to elude these flies. They shiver and shake violently to dislodge the warble flies, or race frantically to escape them. And they stand, wild-eyed and distraught, with muzzles close to the ground, trying to avoid the insistent, pressing botflies. But all to no avail; nearly all caribou are infested with these parasites.

Early fall finally brings relief from summer's insect plagues. The caribou graze in relative peace on the vast and verdant arctic meadows, and they grow sleek and fat, resplendent in their new, clove-brown, short-haired coats. The great tide of life that swept so urgently to the North in spring has run its course; the caribou begin to ebb back towards the forest. The herds now are mixed; the bulls, in prime condition, have joined the cows and calves. In October, when snow covers the land again and ice glazes ponds and lakes, the

RIGHT ABOVE: *The thimble-sized chicks of the Baird's sandpiper resemble bits of tundra lichen to perfection.*

RIGHT BELOW: *In red phalaropes, the sex roles are reversed. The female courts the male, she lays the eggs, and leaves. The male, here, incubates the eggs and lovingly "mothers" the chicks.*

FOLLOWING PAGES: *The mountain slopes of northwest Greenland are speckled with dovekies in late May. More than eighty million of these starling-sized seabirds breed in the Arctic.*

caribou mate, and then continue south into the forest belt, completing the great cycle of their migration.

Unlike the caribou's pelage, which varies little with the seasons, the plumage of the ptarmigan, the pigeon-sized grouse of the North, is in constant harmony with the prevailing colours of the land. Ptarmigan are snowy white in winter and piebald in late spring (part winter's white, part summer's brown). They assume rich earthy hues of brown and grey in summer, become speckled grey, white, and brown in fall, and resume their dazzling white when snow covers the land again. They seem to be aware of their cryptic coloration and favour matching backgrounds. White ptarmigan are extremely loath to cross dark ground, and birds in summer brown avoid patches of snow.

As a result, ptarmigan are inconspicuous at all times of the year except in spring when the cocks ardently court the females. Each cock establishes a territory which he defends furiously against the incursions of other males. Near the female he struts and burps and clucks, the serrated combs above his eyes glowing in brilliant vermilion. With wings akimbo, stiff, and dragging, and his tail fanned so that its black rectrices flash out in startling contrast to the encompassing white, he pursues the female with mincing, urgent steps. He suddenly soars aloft with a sharp clatter of wings, then drifts down towards the female with an excited, gurgling song.

The tundra, for many months so still and white, is suddenly vibrant with colour and life. Meltwater murmurs down to lakes and rivers in a shimmering lacework of brooks and rills. Mosses and lichens, dry and drab in winter, avidly absorb the moisture and swell and shine with renewed life. Tiny rosettes of purple saxifrage dot the sun-warmed ridges. Furry, carmine-tipped catkins rise from the prostrate dwarf willows. Arctic poppies on long slender stems turn their golden blooms towards the sun. Woolly lousewort, wrapped in the gossamer fuzz of myriad silvery hairs, opens its pink blossoms. The tiny bells of arctic white heather nod gently in the breeze, and the deep-purple flowers of Lapland rhododendron glow in marshy hollows.

Like a late-season snowstorm, immense flocks of snow geese arrive at their tundra nesting grounds. In all, they number nearly two million. At one colony alone, in the delta of the McConnell River near the west coast of Hudson Bay, more than a quarter million snow geese breed. The elegant Ross's geese, among the smallest and rarest of North American geese (their total population is slightly more than 100,000), fly from their wintering grounds in California to nesting areas on the northern tundra, regions so remote and inaccessible that the first nests of this dainty little goose were discovered only in 1938. On the ponds and lakes that dot the tundra, loquacious oldsquaw drakes and ducks babble, hoot, and whistle, and near the coast elegant eider drakes court their buxom females with gentle, crooning woodwind notes.

An arctic fox relaxes on a flower-dotted meadow in northern Alaska. Pure white in winter, arctic foxes change into brownish fur in summer.

Spring brings the host of shorebirds to the North. Sanderlings and other sandpipers probe the mud near ponds and brooks for hidden larvae and tiny worms, leaving a filigree of delicate three-toed tracks. Golden plovers swoop in the sky in the intricate gyrations of their courtship flights, and black-bellied plovers perch on tundra knolls and fill the air with their plaintive fluting.

Snow buntings are the first of the migratory birds to come to the North. When they arrive, snow still covers the land, and they flit in nervously twittering flocks from one bare patch of ground to another, picking up last year's seeds. In the Arctic they are the harbingers of spring, and they nest as far north as there is land: on northern Ellesmere Island, northernmost Greenland, on icy Franz Josef Land north of Russia, and on the New Siberian Islands. Soon after they arrive on the tundra, male snow buntings stake out territories, proclaiming possession from elevated perches, a boulder or a knoll, with a tinkling, trilling song. Elegant horned larks spiral towards the sky until they are but specks in the blue, then drift gently downward on set wings with a jubilant, lilting song.

As if roused from their eight-months sleep by this triumphant symphony of spring, ground squirrels emerge from their burrows. Throughout the North they are know as *siksiks*, their onomatopoeic Inuit name, and in Alaska they are called "parka squirrels," for natives there used to make parkas from their pelts. When they retired in September from the hostile upper earth to well-stocked hibernation dens, the squirrels were exceedingly fat. Now, in spring, they are lean and famished, having lost nearly half their weight during winter's long sleep.

Arctic ground squirrels live in scattered colonies. Their favourite habitats are dry slopes and sandy ridges. There, males in spring establish territories, the strongest taking the best terrain. Poorer areas fall to weaker males, who are probably doomed to bachelorhood, since females with a fine sense for the practical choose their mates not for looks but for their property.

In addition to its "home" den, which usually has a maze of tunnels, several chambers, and many exits, a ground squirrel digs in the vicinity many "duck holes," shallow burrows into which it can slip in case of an emergency. A *siksik* leaves its burrow with utmost caution. A dark, twitching nose appears; the head inches up a bit farther; alert, slightly protuberant eyes look worriedly about, and slowly, with low mutterings and many feigned retreats to provoke a hidden enemy into betraying himself, the squirrel finally emerges, sits bolt upright near its burrow, and inspects the countryside long and carefully. According to a scientific report, ground squirrels spend 14 per cent of their active time in observation.

The squirrels rely on such precautions, on speedy flight, and, above all, on their excellent alarm system to protect them from their many enemies. About once a minute a foraging squirrel sits upright and looks carefully around. The

A glaucous gull on cloud-wreathed Prince Leopold Island in Canada's Arctic, where these powerful birds eat primarily murre eggs and chicks.

moment it sees danger, it calls a warning, and this alert is instantly picked up and passed on by all its neighbours. These warning cries are quite specific: a shrill whistle means a bird of prey is approaching; a raucous, scolding call signifies the arrival of a terrestrial enemy. The closer the foe, the louder the squirrel screams. As a fox or wolf passes through *siksik* area, its progress is pinpointed and passed on by a sound-wave of warning calls.

Despite such alarms and mutual warnings, most squirrels' lives are short. They are a favourite food of rough-legged hawks and snowy owls; in the western Canadian Arctic and parts of Alaska, *siksiks* are the main prey of golden eagles; gyrfalcon, gliding low across the tundra, snatch them up; Barren Ground grizzlies dig out entire colonies; and the slinky ermine slithers into the labyrinthine tunnels and kills them in their dens. The wariest squirrels, with luck, survive. The females mate, clean out natal chambers, evict the males, build cozy nests, and in them, after a twenty-five-day gestation period, the tiny, naked young are born.

Spring in the North is but a fleeting, joyous interlude between the white, austere winter and the throbbing urgency of an all-too-brief summer, a time of awakening and of wonder, when all nature's pent-up energy and vitality bursts forth in passionate renewal, in resurrection, and in life.

Widespread and adaptable, herring gulls nest from the high Arctic as far south as Mexico. While its parents forage for food, a sleepy chick opens its beak in a prodigious yawn.

The Birds of Summer

IN the Arctic, said the explorer-writer Peter Freuchen laconically and with some truth, "July is summer." In the Far North this is the only month whose average temperature is above the freezing point. June is fair but often fickle; in August summer's brief glow is rapidly fading. In early July, the tundra is spangled with flowers. A month later, the tiny petals have paled, shriveled, and dropped.

Because summer is so desperately short, the fervid exaltation of spring is glorious but brief. There is no time for leisurely dalliance, for all must try to raise a new generation within summer's brief span. In May, young ravens are already hatched. Lemming mothers nurse their second or third litter of the year. In clefts and niches among the rocks, snow buntings incubate their eggs in deep-cupped grass nests, warmly lined with hairs and feathers.

On off-shore holms and skerries, eider ducks brood their large, mat-olive eggs in nests cozily padded with greyish eider down. These colonies are often dense. When the explorer Elisha K. Kane visited Littleton Island off northwest Greenland in the early summer of 1854, there were so many ducks "that we could hardly walk without treading on a nest."

Some birds in warmer lands raise two or more broods during summer's long sway. For birds nesting in the Arctic, raising one brood is a race against time, for their young must be ready to fly and flee before the onset of winter. While this precludes sequential broods, some sanderlings, members of the sandpiper family that nest in the arctic regions of both America and Eurasia, increase their progeny by raising two broods simultaneously. The female lays a clutch of four eggs and turns them over to her mate to brood and raise, then lays another clutch and raises it herself.

The sanderling, said naturalist John K. Terres, is a cosmopolitan globe-trotter, "few species equal its worldwide wanderings," and most of the ones that do also nest in the Arctic. Pale grey in winter, sanderlings probe the sea-washed beaches of southern Chile for tiny crustaceans, mollusks, and worms. In March they leave their South American winter haunts and, after a flight of about 8,000 miles, reach their tundra nesting grounds in late May or early June. The sanderlings of the Eurasian Arctic winter in southern Africa, in Australia, and New Zealand.

Less than ten bird species are year-round residents in the Arctic: ptarmigan, raven, snowy owl, gyrfalcon, some hoary redpolls, two species of gull, black guillemots, and a few eiders. Most "arctic" birds are migrants, fair-weather

Oldsquaws are among the most common and noisy ducks of the North. A male rests upon a boulder in early summer at the edge of a partly frozen tundra lake.

ABOVE: *Bald eagles normally nest in tall trees, but on Alaska's treeless tundra they raise their young upon the ground.*

RIGHT: *A young snowy owl on the tundra. In lemming-rich years, snowy owls raise many young. When the lemmings fail, most owlets die.*

ABOVE: *A parasitic jaeger rushes furiously at an intruder, valiantly defending its nest. Jaegers can catch their own food but, whenever possible, prefer to rob gulls and terns.*

RIGHT: *A regal peregrine falcon gazes out over the tundra. The peregrine hunts birds in flight, swooping on them at speeds that can attain 200 miles per hour.*

visitors that come to breed in summer when the food wealth of the North attains a brief peak, and leave hastily in fall before the onslaught of winter, some spanning continents in vast and complex migrations that are miracles of precision navigation and endurance.

The elegant knot, a stocky shorebird with rust-red face, chest, and belly, breeds around the top of the world, and in Canada and Greenland as far north as there is land. The knots arrive on northern Ellesmere Island in late May and early June, when nearly all the ground is still covered with snow and ice and the earth is frozen as hard as rock, and fill this bleak and icy desolation with the gentle, mellow oboe notes of their courtship song.

The moment the snow leaves the ground, the knots nest. Their chicks hatch around mid-July when insects are most abundant, grow extremely rapidly, and are fledged by early August. The adults leave, followed about two weeks later by the young birds. The knots from northern Ellesmere Island fly across northernmost Greenland, those from southern Ellesmere and Baffin Island cross Baffin Bay and Greenland's ice cap, and all the knots, including those from Greenland, in all about half a million, fly to Iceland to rest and feed, and then continue to Britain and Holland, where they spend the winter.

The knots from the lower and the western Arctic follow a completely different route. They fly across Canada and the eastern United States, then 2,000 miles over the sea to Suriname, cross Brazil and spend winter along the eastern coast of Argentina, as far south as Tierra del Fuego. The knots from arctic Europe fly to Africa; those that nest in Siberia winter in Australia and New Zealand.

Golden plovers accomplish a similar odyssey. The birds from arctic Canada and most of Alaska fly to the Labrador coast, rest there a while to feed and fatten, and then fly non-stop 2,800 miles across the sea to Suriname and on to their winter quarters on the Argentinean pampas. The golden plovers of western Alaska set out boldly across the Pacific and arrive forty-eight hours and a quarter of a million wing beats later in Hawaii and from there many fly another 2,100 miles to the Marquesas Islands and beyond.

Even small birds make immense migrations. The slender little arctic warbler weighing barely a third of an ounce flies from its breeding grounds in western Alaska to the Philippines and to Indonesia. The yearly travels of the wheatears are even more impressive. These perky, ash-grey, dark-winged birds of the thrush family prefer the stony barrens, where they build their nests in clefts among the rocks. When they bob excitedly, or suddenly fly up, their white rumps flash brilliantly. To this the wheatears owe their name, a bowdlerized version of the original "white arse." Like the arctic warbler, the wheatear is essentially an Old World species and it has invaded arctic America both from the east and the west. In fall, obedient to ancient promptings and patterns, nearly all wheatears retrace ancestral migration routes to central Africa, the

Ethereally beautiful yet incredibly hardy, the pure-white ivory gull, the "phantom of the polar ice," nests only in the farthest North.

A common eider duck relies upon camouflage plumage to avoid detection. When she leaves the nest, she covers her olive-green eggs with eider down, to hide them and to keep them warm.

A resplendently coloured king eider drake displays his finery on a tundra pond on Canada's northern Devon Island.

ABOVE: *A young rough-legged hawk upon its tundra nest. Lemmings are its main food. In years of lemming scarcity, rough-legged hawks may not breed at all.*

RIGHT: *Its nest hidden in a clump of sedges, a short-billed dowitcher stands warily nearby in the shimmering grass of an island off Alaska.*

ones from Alaska crossing all of Asia, those from eastern Canada flying via Greenland, Europe, and North Africa. A few wheatears, however, have lately discovered a less arduous alternative and now spend their winters in Florida, Bermuda, and Cuba.

The champion migrant of all birds is the arctic tern, which yearly and with seemingly nonchalant ease commutes between the antipodes. Arctic terns nest in the North to the very limits of land and fly in fall from the highest Arctic to the margins of Antarctica, a return trip of about 25,000 miles. If one adds to this their daily flights while foraging in the North and the South, they probably cover more than 30,000 miles each year. One banded bird caught in 1970 was thirty-four years old. In its lifetime this delicate, four-ounce tern may have flown well over a million miles.

Arctic terns are irascible, spunky, shrill-voiced birds that usually nest in colonies. They argue a lot with neighbours, but the moment an enemy approaches, all soar aloft and attack united and with fearless fury. I once watched a polar bear amble leisurely towards a small tern colony on northern Somerset Island. The terns dived like glittering arrows and hit the bear with scarlet stiletto beaks. Annoyed, the bear shook his head, snapped futilely at his elusive tormentors, walked faster, and was finally driven from the terns' territory. The birds returned to the nests, shallow scrapes in the gravel, and settled upon their speckled eggs with a gentle, shuffling motion. Off-duty birds flew out over shimmering lagoons, hovered above the water, plunged, and emerged with small fish.

Superb flyers that the terns are, one arctic bird can fly circles around them. Parasitic jaegers, sleek and swift, with dusky plumage, rakish black caps, and long, hooked bills, are expert plunderers. These gull-like birds with hawk-like habits are quite capable of hunting themselves, but they much prefer to harry hapless terns and gulls returning from the sea with food. They chivvy them so mercilessly that they finally drop or disgorge their catch, which these aerial pirates snatch up with a graceful swoop.

Of the three species of jaeger that nest on the tundras of the North, the long-tailed jaeger is the smallest and least piratical. It catches lemmings, small birds and their young, and eats large amounts of insects. While one mate incubates, the other usually stands on guard nearby, ready to attack and chase any enemy that comes into the vicinity. The jaegers, too, are long-range migrants. Most spend the winter off the coasts of South America.

Among the last of the great migrant host to arrive in the North, and among the first to leave, are the beautiful phalaropes, robin-sized shorebirds with very peculiar habits. Their feet are lobed, and their contour feathers cover a dense layer of down which imprisons warm air near the skin. They are the most aquatic of all waders, spending winters at sea off South America. Phalaropes come to their tundra breeding grounds in early summer and twirl buoyantly

Red-throated loon upon its nest at the edge of a tundra lake. Good swimmers and expert divers, the loons are barely able to move on land.

upon lakes and ponds. They spin like tops upon the water (one observer saw a phalarope make 247 consecutive revolutions) and pick with rapid, jerky dabs small organisms from the roiled water. The superabundant mosquito larvae and pupae are one of their principal foods.

In phalaropes sex roles are reversed. The female arrives in the North in a resplendent summer plumage of rufous red, rich browns, tan, and white. The male is more modestly attired. The female courts the male. She sings and displays in his presence and turns in furious jealousy upon any rival female that dares to come near her chosen male. ''All the fights we saw were between females,'' wrote the ethologist Niko Tinbergen, who watched courting phalaropes in East Greenland. Both male and female build several nests, but it is the female who finally selects one, lays her eggs, and leaves them in the care of the male. He incubates the eggs and ''mothers'' the chicks. Some females are polyandrous. They mate with several males, provide each one with a clutch of eggs, and leave them to their tasks. The females take no further interest in either mates or offspring and, freed from all parental responsibilities, fly soon afterwards to the South.

In fall, when all migratory birds head south to escape the approaching winter, one species flies towards the north. Small flocks of Ross's gulls pass Point Barrow, the north tip of Alaska, in October and then vanish northeastward into the polar night. While drifting with his ship *Fram* in the pack ice of the Arctic Ocean (1893-1896), the Norwegian explorer Fridtjof Nansen saw several of these delicately rose-hued gulls and was enchanted by ''this rare and mysterious inhabitant of the unknown north . . . of which no one knows whence it cometh or whither it goeth, which belongs exclusively to the world to which the imagination aspires.''

In 1905, Russian scientists found the first colony of Ross's gulls breeding on swampy tundra not far from the taiga's edge, and several other colonies have since been located in the same region of northeastern Siberia. In the 1970s, a tiny colony was discovered on a small, remote island in the high Arctic of Canada, and in the summer of 1980, three pairs of Ross's gulls suddenly appeared on the west coast of Hudson Bay. They built their nests two miles from the town of Churchill, and only a few hundred feet from a highway, and returned to the same location in subsequent summers.

In fall, these dainty little gulls head north, and it is assumed that they spend the winter upon the polar pack, that region which Freuchen called ''the most horrible part of the world, the ice desert near the Pole.'' There, in total darkness, lashing storms, and deadly cold, they somehow survive by finding enough food in the leads and pools of open water between the ever-shifting ice.

Nearly as beautiful as the Ross's gull is another bird of the farthest North, that ''phantom of the polar ice,'' the pure-white ivory gull. It nests only in the highest Arctic and is probably the only bird to have ever nested at sea, on ice

Hardy, frugal, and astute, the raven is one of the few birds able to spend both summer and winter in the Arctic. Of all northern birds, it is the first to breed in spring.

floes covered with morainal detritus. Most of their breeding colonies are in extremely remote and inaccessible places, occasionally inland on nunataks, bare mountains surrounded by glacial ice, and often far from the sea. But there, on steep cliffs hemmed by ice, they are safe from arctic foxes and polar bears that have been known to raid more easily reached colonies.

Ivory gulls are birds of the ice; they rarely stray far away from it. In summer they catch fish and invertebrates at leads among the floes, patrol beaches, gorge on carrion when they find it, and occasionally kill young birds. In winter they roam the infinite pack ice, in darkness and in bitter cold. They feed at leads where dead fish and crustaceans sometimes float to the surface; they trail polar bears in hope of scraps; they eat the faeces of bears and foxes; and somehow, nearly miraculously, they manage to survive in this frozen, pallid world of utter desolation.

The migrants avoid such hardships. In this land of feast or famine, they partake only of its short season of plenty. They arrive in their legions when spring wakes the North. The hatching of their young coincides with the greatest abundance of the food they require. All feed voraciously, to grow and to accumulate the fat reserves that will sustain them and give them strength during their long flight to the South, away from this land that in the few, intense weeks of summer knew the fullness of life.

A common eider duck incubates her eggs. Her nest is lined with warm eider down plucked from her breast and abdomen.

Feast Before Famine

FALL comes early to the Arctic. In August, the nights lengthen; snow dusts the mountains; and in the mornings small pools are glazed with a film of delicately veined ice. Hairy bumblebees, torpid with cold, wait for the sun to warm them so they can fly again and visit late-blooming flowers.

The tundra is spattered with the vivid colours of fall. The leaves of arctic willow turn golden yellow; the tiny, serrated leaves of dwarf birch are brick-red; the waxy, reticulated bearberry leaves shimmer in brilliant crimson in early fall, then turn a rich, sombre wine-red. The silver-glistening bolls of arctic cotton grass, so pert and round in summer, are frayed into gossamer threads by the tearing winds of fall.

The berries are ripe: blue bilberries, juicy and sweet; the watery, sourish crowberries; and the amber-yellow cloudberries that are extremely rich in vitamin C. They are a favourite food of lemmings, of arctic ground squirrels, and of their arch-enemy, the Barren Ground grizzly. Jaegers pick berries; arctic foxes feast on them and for a while their scats are berry-blue; great glaucous gulls waddle awkwardly across the tundra and eat berries; and since at this time of the year there is little else for them to eat, polar bears occasionally wander inland to harvest the abundant berry crop.

Most birds are nearly fledged. Young rough-legged hawks stand on their big stick nests and flex their wings. These hawks prefer to nest on cliff ledges, ideally beneath an overhang that shields them from snow and rain. But in those large areas of rolling tundra where cliffs are rare or non-existent, the hawks build their nests upon the ground, usually on a slope overlooking a valley. On treeless islands off Alaska, bald eagles face a similar problem. Normally, they build their large nests in tall trees. On the islands, the eagles nest on the ground. The number of young a pair of rough-legged hawks raises depends primarily on the stage of the lemming cycle. In years of lemming abundance, the female hawk may lay as many as seven eggs and all the well-fed young fledge. When lemmings lack, the hawks can only raise one or two young.

Snowy owls are nearly as dependent upon lemmings. These white, dusky-flecked owls of the North, with a five-foot wingspread, are powerful hunters. They kill ptarmigan, murres, and eiders, they snatch char and trout from river shallows, and although the snowy owl weighs only about four pounds, it can catch and kill a twelve-pound arctic hare. But lemmings are its staple fare. The presence or absence of lemmings stimulates or inhibits the female owl's fecundity. In years of lemming scarcity, she lays only two or three eggs, or may not

Kittiwakes, small, noisy northern gulls, nest on narrow ledges and cornices of soaring cliffs.

breed at all. In peak lemming years, the owl's clutch is large: as many as fifteen eggs have been recorded.

The female snowy owl begins to incubate as soon as the first egg is laid, and the young hatch at intervals; the last may emerge two weeks after the eldest. These late-comers are often doomed; the older owlets trample them in their eagerness to obtain food. The young that die are usually eaten by the mother or she may feed them to their stronger, more fortunate siblings.

After three weeks, the owlets leave the nest and scatter, grey downy gnomes with bright yellow eyes. They wheeze loudly and insistently whenever they spot a parent carrying food. The owlets seem forever hungry. With winter only weeks away, there is an intense urgency in the young to feed, to grow, and to become independent from parental care.

In fall all arctic life prepares for winter. Urged on by special hormones, migratory birds feed nearly incessantly to accumulate the fat reserves that will fuel their long flight to the South. Ground squirrels, who have nearly doubled their weight since spring, busily stock winter dens with food supplies. Barren Ground grizzlies are swathed in fat so thick it will sustain them until spring. Arctic foxes cache surplus food, lemmings, or young birds, to ease winter's hardship. Char, the large, far-northern trout, who have feasted all summer at sea, return in fall to the lakes where they will spawn. Inuit once speared them in complex stone weirs called *sapotit*, and cached large quantities of the fat fish as food for winter. The caribou are sleek in fall; about a fifth their total weight is fat. Even plants store lipids in their roots and rhizomes for that magic moment, nine months hence, when winter will yield again to spring, and instant, rapid growth will be imperative.

Insects can neither flee from the arctic winter, like birds, nor can they shield themselves like mammals from its lethal sting with thick layers of fat and fur. Instead they produce within their bodies a glycerol-like antifreeze so efficient they can survive, unharmed, temperatures of $-70°$ F. They spend winter in suspended animation and resume life nine to ten months later when the warmth of spring awakes them.

The birds become more and more agitated. Increasing cold, shortening days, and declining food trigger "Zugunruhe," as ornithologists call the nearly febrile restlessness that grips them just prior to migration. Dense groups of knots zigzag into the sky, flashing alternately in red and brown as all birds of the flock turn abruptly and in unison. Oldsquaws gather in large rafts and swim, neatly spaced and aligned, back and forth along the coast. Nervous, twittering flocks of snow buntings drift like snowflakes across the land. Skeins of snow geese wing purposefully towards the South. Each day more birds depart, and snow and silence descend again upon the North.

Ptarmigan, ermine, and foxes change into winter white, as do the hares of the lower Arctic who were an earthy brownish-grey in summer. In the Far

---

RIGHT ABOVE: *Short flippers and a fat, fusiform body make it difficult for a harbour seal to scratch an itchy spot during the annual moult.*

RIGHT BELOW: *A female harbour seal plays with her flipper. These much-hunted seals breed on isolated islands and remote beaches.*

North, where summer lasts barely two months, arctic hares remain white all year. They also have some very peculiar habits not shared by their more southerly cousins. In late fall and winter, they gather in large flocks. On Ellesmere Island herds of thousands of hares have been seen. When alarmed, these far-northern hares stand bolt upright and, as the biologist David Gray observed, even bounce "up and down on tiptoe while assessing the danger," and then bound away on long hind legs in rapid kangaroo leaps.

These hares of Canada's high arctic islands are, as a rule, extremely wary in summer when, white upon a dark land, they are startlingly conspicuous. And they can be amazingly tame in winter, when their white fur blends perfectly with the snow. The explorer Otto Sverdrup once found himself surrounded by a group of arctic hares so trusting he could pet them, which made him feel like "Adam in Paradise before Eve came." The hares were not even scared of his dog teams: "There were such legions of them and they scurried about so in all directions," his dogs became frantic with excitement. The hares "did not appear to be afraid; they hopped about only a few yards in front of the teams. . . . As if to incite the dogs to the utmost, the hares came and settled down a few yards from them, and then stood on two legs and stared at us."

In fall, these high-arctic hares cluster, usually in herds of 100 to 200 animals, occasionally in immense flocks, hordes of hares that move ghostlike through the polar night. In 1971 on western Ellesmere Island, a Canadian Wildlife Service biologist saw groups of thousands, and estimated that there were 25,000 arctic hares within an area of less than five square miles.

Arctic hares eat grasses, sedges, leaves, and flowers in summer, and in winter mainly arctic willow. These large hares, which weigh about twelve pounds, are superbly insulated from the severe cold (the mean January temperature on Ellesmere Island is − 30° F.) by a thick layer of dense, down-soft underwool beneath long, silky fur. They are careful to husband both warmth and energy. After feeding, David Gray observed, the hares tuck their short tails between their hind legs, press the forepaws into the chest fur, lower the ears into the fur on their back "and settle into an almost perfect spherical shape with only the thick yellowish [densely furred] pads of their hind feet touching the ground."

In spring, the hares go mad. Sverdrup observed their zany antics near a bay on southern Ellesmere (which he named "Hare Fiord") and concluded they had "lost their heads from love." One hare that David Gray watched streaked across the tundra with wild, erratic leaps that "propelled him straight into the air a good metre and a half," hit the ground with racing feet, and zoomed off into another *grand jetée*.

Two months after these exuberant spring revels the young are born, usually four or five, occasionally as many as eight. The leverets, as young hares are called, are densely furred in grey-flecked brown that blends to perfection with the tundra. They are nearly impossible to spot until a female appears near them

Surrounded by glaucous-winged gulls hoping for scraps, a Kodiak bear eats a salmon caught in river rapids.

and growls, her signal to the hidden leverets to come and nurse. Only then, noted the biologist G.R. Parker on Axel Heiberg Island, "did the abundance of young become evident. At such times the ground seemed to explode with young hares rushing to meet the females."

When startled or pursued, arctic hares nearly invariably head uphill. On level ground, their main enemies, wolves and arctic foxes, may equal their speed, but healthy hares probably always win an uphill race. A wolf pack, though, usually hunts as a skilful, highly coordinated team and often manages to outmanoeuvre a fleeing hare. But hares, too, can be wily. One, chased by a fox, sought refuge near humans and stayed within ten feet of them "until the fox had given up and departed." David Gray once also saw three hares that were hunted by wolves run to a nearby muskox herd and remain close to these massive "protectors" until the wolves left.

Of all the North's wildlife, muskoxen and polar bears made the greatest impression on arctic explorers. "We see, we talk, and we dream more of bears than of any other animal," wrote Captain Leopold M'Clintock near the end of a two-year (1857-59) sojourn in the Far North. And of the muskox he said, "Nothing in these dreary solitudes brings home to one so forcibly the wonderful power of adaptation . . . as the first sight of a herd of these tremendously shaggy little buffaloes, contentedly scraping away the snow, and browsing such scanty vegetation as the soil affords."

In this, M'Clintock was only partially right. Muskoxen are certainly shaggy and they are marvellously adapted to the Arctic. But they are not little and they are not buffaloes. A mature bull weighs about 800 pounds (one, in captivity, reached 1,400 pounds), and despite their scientific name *Ovibos moschatus* – the musky sheep ox – muskoxen are not ovine or bovine, and only bulls in rut have a musky smell. They belong to the ancient family of goat-antelopes and their nearest relative is the rare takin of Tibet.

The explorers' fascination with these animals had a ruthlessly practical side. Muskoxen were very easy to kill. M'Clintock and two companions shot an entire herd "in three or four minutes." Apart from man, the wolf is the muskox's main enemy. On rare occasions, a polar bear may kill a muskox. During an aerial game survey in 1981, the biologist Anne Gunn came upon a Barren Ground grizzly that had just killed a solitary muskox bull after what seemed to have been, to judge by the torn, blood-spattered tundra, a fierce and prolonged struggle. But only wolves are a serious threat to muskoxen, and against them they have evolved an extremely effective form of defence. When confronted by wolves, the muskoxen of a herd bunch into a compact circle or semi-circle, sharp-horned heads turned towards the enemy. Within this furry bulwark, calves are hidden and protected, tightly pressed against their mothers' flanks. From this defensive circle, adults make lightning-quick sorties against the wolves, try to gore them, and then return into the circle. It is a superb and

RIGHT TOP: *A male hooded seal at rest, the hood or crest on top of its head deflated.*

RIGHT MIDDLE: *An adult hooded seal male with inflated hood. Hooded seals sometimes blow up this elastic skin crest when they are annoyed, but often do it for no apparent reason at all.*

RIGHT BOTTOM: *In addition to inflating its hood, a hooded seal male can also blow up its nasal septum into a brownish or orange balloon.*

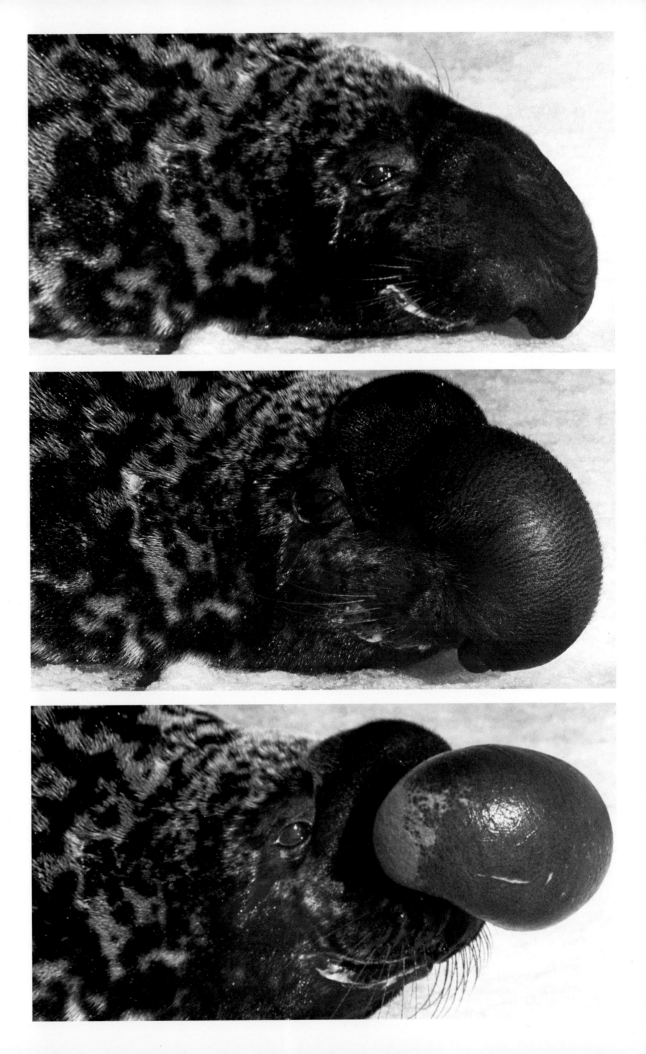

usually successful stratagem against wolves; it was fairly efficient against Inuit armed only with lances; but it was suicidal when men came with guns. The expeditions of Robert Peary and Otto Sverdrup alone killed more than 1,000 muskoxen on Ellesmere Island. Although at times desperately in need of their meat, Sverdrup was sickened by the slaughter of these "defenceless animals which had set themselves up as targets."

Whalers and gun-armed Inuit killed tens of thousands more, and in 1917, when Canada's muskoxen received total protection, only about 500 were left on the arctic mainland, and a few thousand on the immensity of the far-northern islands. Since then they have increased, and muskoxen from East Greenland and Canada have been introduced to regions where they had long ceased to exist: to Spitsbergen, Norway, Alaska, arctic Quebec, and Siberia.

Thanks to decades of protection, Canada's muskoxen now number about 40,000 and Inuit may hunt them again on a quota basis. In Alaska and Canada limited sports hunting is now also permitted, although, as the famous explorer Vilhjalmur Stefansson once remarked, "the word 'sport' has a curious meaning when applied to killing muskox. . . . I would say that equally good sport could be secured with far less trouble and expense by paying some farmer for permission of going into his pasture and killing his cows."

When not harried by hunters or by wolves, muskoxen lead a placid, unhurried existence, alternately feeding and resting. Some live in regions so hostile and with such sparse plant growth that it seems incredible these large animals can survive there. In northern Ellesmere Island, where muskoxen are numerous, the arctic winter night lasts four and a half months. From the middle of October to the end of April the temperature is below zero degrees F. For about seventy days it is below $-50°$ F. and it can drop to $-70°$ F. Storms are frequent, those lethal arctic storms that carry with them, as Peary said, "a roaring, hissing, suffocating Niagara of snow." But, superbly protected by thick wool and fur so long it hangs nearly to the ground, muskoxen stolidly endure this deadly climate.

Muskox calves are nearly as hardy as the adults. They are born in April or May, a time of lashing storms and temperatures that can drop to $-30°$ F. Emerging from the steady and cozy $101°$ F. of its mother's womb, the newborn calf must survive and adjust to a 130-degree temperature drop. Most do; calf mortality seems to be low. Wrapped in dark, dense, curly wool, steaming at first in the icy air, the calf staggers to its feet a few minutes after birth and begins to nurse. Within a week it supplements its mother's rich milk (which contains 11 per cent butterfat) by browsing alongside her on the scanty vegetation. It remains with its massive, protective mother for at least a year. This contributes greatly to the calf's chances for survival. But as a result, females bear calves, at best, only every second year.

The even, placid tenor of the muskox's life changes abruptly in fall. Then the most powerful bulls claim possession of the herds, expelling all other adult males. Some wander off, resigned, it seems, to bachelorhood, but most remain near the herds, a constant threat to the herd bull's primacy. When a rival comes too close, the lead bull charges. Hair-skirts flying, the great animals thunder across the tundra and, despite their bulky, ponderous appearance, twist and wheel with the speed and dexterity of a quarterhorse.

If a rival does not flee, the protagonists square off for a battle as ritualized as a mediaeval jousting match. They face each other, slowly back up, shaking their great, horned heads, stop when they are about 150 feet apart, and suddenly, as upon command, both charge at top speed and collide, boss against massive boss, with a dull thud that can be heard more than a mile away. They push and hook, then back up and charge again – and again and again, until one contestant reels after impact, wheels, and flees, briefly pursued by the winner.

By September, the battles of the bulls have ended. The winners have mated with the oestrous cows. The outcast males return to the herds. Fall's feast is nearly over and all feed in peace and amity, now intent only upon accumulating fat reserves against the rigours of the nearing winter, when all life will be tested to the utmost in a silent, frozen land.

The Bounteous But Imperilled Seas

THE animal wealth of the arctic lands is great; that of the polar seas is infinitely greater. The arctic and antarctic seas are stupendously rich in nutrients, richer than all other seas on earth, and hence are home to hundreds of millions of seabirds, to millions of seals, and to whales, the largest creatures this world has ever known.

They all require prodigious quantities of food. A single whale gulps down two to three tons of food each day. The northern fur seals eat three and a half million tons of fish each year. The walruses require roughly 10,000 tons of clams per day. And the murres of the arctic and sub-arctic regions eat about six billion pounds of fish each year. These animals are at or near the apex of a food pyramid whose plant base is of astounding richness, for one whale consumes via its food chain about five trillion sea plants a day.

One of the first to grasp the enormous productivity of the northern seas was the Englishman William Scoresby (1789-1857). Like his father, he was a highly respected and successful whaling captain. He was also a famous explorer, an eminent, though self-taught scientist, and he ended his days as an Anglican minister. His interest in the sea's minute plants and animals (its phytoplankton and zooplankton) had a practical aspect. Wherever in early summer these tiny beings were so numerous as to colour the sea a deep brown or olive green, he noted, there was a good ''probability of finding whales.''

Scoresby examined one drop of this life-coloured water under a microscope and found in it ''about 26,450 animacules.'' From this he calculated that a body of water two miles square and 250 fathoms deep would contain about 23,888,000,000,000,000 (23 quadrillion) of these ''animacules.'' He then contemplated the amount of life that must inhabit the vastness of the polar seas (which cover about 5,650,000 square miles) and exclaimed in wonder: ''what a stupendous idea this . . . gives of the immensity of creation!''

If Scoresby was awed by the abundance of life in the northern seas, he was also aware of its frailty and interdependence. From diatoms so tiny he could see them only with the aid of his microscope, to the giant whales he pursued, there existed in the sea, he wrote, ''a dependent chain of animal life, one particular link of which being destroyed, the whole must necessarily perish.''

A century and a half after Scoresby meditated upon the wealth and vulnerability of the arctic seas, Prime Minister Pierre Elliott Trudeau spoke about the same subject to journalists in Toronto. The ''continued existence in unspoiled form'' of the polar sea, he said, ''is vital to all mankind. The single most

Only one shark species inhabits the northern seas, the large, lethargic Greenland shark. It was formerly extensively hunted for its huge, vitamin-rich liver.

imminent threat to the Arctic . . . is the threat of a large oil spill.'' Such a disaster in the Arctic, said Trudeau, could affect ''the quality, and perhaps the continued existence, of human and animal life in vast regions of North America and elsewhere.''

The oceanographer Jacques-Yves Cousteau pursued this possibility to its stark conclusion in his apocalyptic vision of our fate if we destroy life in all the seas by dumping lethal substances into them. Then the global climate will change abruptly, the air's oxygen content will decrease sharply, and thirty to fifty years after the oceans have died, all mankind (and all animals) will perish ''in unutterable agony.''

Despite their seeming isolation, the polar seas are part of a worldwide web of moving water. It is this very interchange of water that is, in part, responsible for the stunning animal wealth of the northern seas. Such mighty ocean streams as the Kuroshio and its prolongation, the North Pacific Current, and the Gulf Stream and its extension, the North Atlantic Current, carry warm water from the tropics far to the north, while the Labrador Current and the East Greenland Current pour icy, arctic water towards the south at the rate of four million cubic metres per second.

Wherever such currents of differing temperature and salinity meet and mix, they create zones of titanic marine turbulence. The violent, churning currents, the sinking and rising layers of cold and warm water produce immense upwellings from the deep that sweep mineral-laden waters to the surface, abundant nutrients for the minute plants of the sea.

Added to the sweeping, swirling action of the currents is the North's seasonal exchange of water layers. Winter-cooled surface water sinks and is replaced by warmer, lighter, richer water layers from beneath. The vast masses of dissolving ice also contribute to the wealth of the polar seas. Water recently melted from ice contains high concentrations of trihydrol, polymerized water molecules that greatly favour the growth and division of plant cells. Marine productivity, both benthic and pelagic, tends to be greatest in the relatively shallow seas covering continental shelves, and nowhere on earth are these continental shelves more extensive than in the North. The Barents Sea shelf is 750 miles wide, and the continental shelf extends more than 900 miles beyond parts of Siberia's coast.

All these factors combine to produce a sea of nearly boundless fertility. Most numerous and important of the vast number of tiny plants that thrive in these mineral-rich waters are the diatoms, unicellular algae encased in jewel-like silica shells. Each gallon of sea water near the surface contains millions of diatoms. During winter's darkness their productivity is relatively low. But diatoms can maintain photosynthesis, growth, and division at very low light levels, and as soon as the sun rises again above the northern seas and ice, the diatoms grow

---

The unicorn of the northern seas, the male narwhal has a twisted, tapered ivory tusk that can be up to ten feet long. Tusk size may determine hierarchical standing among narwhal males.

rapidly and divide, until they fill the sea like motes of living dust. In early summer, the arctic seas are "in bloom," replete with a plant broth so rich as to render the water nearly opaque beneath the surface and tint its surface brown or green.

Minute planktonic animals browse upon these infinitely rich pastures of the sea, rice-sized copepods, as many as 3,000 per cubic foot of water (each one ingesting 130,000 diatoms a day), the larvae of clams, sponges, and crabs, pteropods, dark, winged pelagic snails that the Polar Inuit call *tulugarssaq*, the ones that look like ravens, and a multitude of euphausids and other crustaceans, collectively known as krill. They are the food of some seals, ringed seals and young harp seals, of many small fishes, herring, capélin, and polar cod, of giant bowhead whales, and of some of the most numerous of the northern seabirds, the dovekies (about 80 million) and the least auklets (about 100 million), and of other auklets and murrelets, of fulmars and of petrels.

From these upper marine layers teeming with life, a steady rain of organic detritus, the dying and the dead, descends to the ocean floor, to be eaten by benthic animals, whelks, worms, crabs, or clams, which are in turn the food of such bottom feeders as walruses and bearded seals. Finally, bacteria within the bottom ooze convert organic wastes into minerals that are swept upwards again by current action and the exchange of water layers. These minerals provide nutrients for myriad minute plants, thus completing the eternal cycle of life and death and life renewed.

This great marine wealth of the North deeply impressed arctic explorers. When John Ross sailed in 1818 into Greenland's Melville Bay, he saw "myriads . . . of little awk [dovekies] . . . together with vast numbers of whales and sea-unicorns [narwhal]." It was this sea wealth that enabled man to live in the Arctic. It had initially lured the Inuit's forbears to the North and had sustained their descendants for thousands of years in this hostile region, for the Inuit, with the exception of a few inland groups who lived mainly off caribou, were essentially sea mammal hunters.

But their numbers were small and they lived scattered over an immense area. They were not nature's masters but subject to her laws. They were highly skilled predators, and like all predators they could not decimate their prey without disastrous consequences to themselves. If in one area a vital prey species declined because too many men had hunted it too successfully, then their hunting success declined and the people starved until the balance between predator and prey had been re-established.

The whalers, sealers, and walrus hunters who followed the explorers, or occasionally preceded them, were not bound by such natural laws and limitations. They did not wish to live in the North; they only came to exploit its wealth, to carry vast quantities of valuable sea mammal products to southern markets. They ceased because of changes in demand (around 1905, petroleum

Walruses sprawl on an Alaskan beach. On warm days they spread their large flippers to dissipate body heat.

began to replace whale oil and featherbone and the newly invented spring steel supplanted baleen, whose price fell from $5 a pound to 40 cents) and because their main prey, whales and walruses, had become so rare that the hunts were no longer profitable. Of an estimated 300,000 Pacific walruses only about 40,000 were left, and probably less than 2,000 bowhead whales remained in the immensity of all the polar seas.

The great commercial hunts of northern sea mammals have ended. Hunting in the Arctic is now nearly entirely restricted to the natives of the North and their take of some sea mammals, such as bowhead, walruses, and narwhal, is limited by quotas. The high price of ivory does lead to infractions. In 1981, agents of the United States Fish and Wildlife Service, posing as investors, bought black-market walrus ivory from dealers in New Orleans. The transactions were tape-recorded and led to raids by eighty agents in five states, from Alaska to New Jersey, and the seizure of 10,000 pounds of walrus ivory worth about $500,000.

Despite such allurements to exceed quotas, hunting does not at present endanger the survival of northern sea mammals. Pacific walruses have increased to at least 150,000 (some scientists believe they may now number 250,000); bowhead whales have increased slightly, to about 3,000; narwhal number at least 20,000, but about 2,000 are killed annually in Greenland and the Canadian Arctic; white whales are still extensively hunted in the seas north of the Soviet Union (and by Inuit in Alaska, Canada, and Greenland), but their circumpolar populations are believed to exceed 100,000. None of the North's seal species is seriously menaced by hunting.

Of the dangers threatening northern marine life, hunting is the one that can be most easily curbed and controlled. The others – the lethal or debilitating effects of toxic wastes, oil pollution, and overfishing – will be infinitely more difficult to control or correct.

The Arctic is a world apart with animals superbly adapted to its harsh climatic conditions. But it is also very much part of a world that is being slowly, insidiously poisoned by the creeping cancer of chemical pollution spreading through the oceans and the air. Although many countries now ban the use of DDT, vast amounts of DDE, its dangerous, long-lived breakdown product, have worked their way into every part of land and sea, and into the tissues of most living creatures. According to a report by David B. Peakall of Cornell University "it is estimated that there are a billion pounds of the substance [DDE] in the world ecosystem." It has been found in considerable concentrations in the fat of arctic polar bears and antarctic penguins. Alaska's Aleuts may no longer eat their favourite delicacy, fur seal liver, for it now contains mercury concentrations far in excess of levels deemed safe by the United States Food and Drug Administration.

Like a glistening mermaid, a female northern fur seal perches upon a wave-washed rock.

Birds, especially seabirds and raptors, both at the apex of food pyramids, are particularly sensitive to such toxic substances. High DDE levels cause peregrine falcons, bald eagles, pelicans, and ospreys to lay thin-shelled eggs that do not hatch. Organo-chlorine residues are blamed for the decline of some gannet colonies where, as a result of these poisons, malformed young are born. Heavy metal poisoning threatens New England's terns. Cormorants are especially susceptible to such noxious substances, and several colonies of double-crested cormorants have disappeared. The Arctic is not immune from this global threat that eventually attains and taints all life, even in the remotest regions of the North.

While chemical poisons seep slowly through the seas and corrode life stealthily, nearly imperceptibly, the results of overfishing are instant and dramatic. The capelin's fate may be a tragic augury of things to come.

Capelin are small (about seven inches long). These silver-sided fish with greenish backs were once incredibly abundant. When capelin are three to four years old, they migrate from the open sea into shoal coastal water to spawn during the highest tides of June. The capelin "scull," say Newfoundlanders, for whom this annual shoreward surge of fish was part harvest and part festival. In his book about Newfoundland published in 1869, Captain R.B. McCrea described the arrival of this "stupendous . . . multitude of fish." The people ran into the surf, stood "up to their hips . . . in fish" and scooped them out with buckets. For Newfoundlanders the prolific and abundant capelin were food, fishing bait, and fertilizer for their fields, and they took about 10,000 tons of them each year.

More importantly, from Russia's Murmansk coast to Iceland, Greenland, Labrador, and Newfoundland capelin were the main food of cod and many other fishes, including Atlantic salmon, as well as of many seals, whales, and seabirds. The capelin, noted a Canadian government report, "are undoubtedly the most important . . . [food] resource in the Canadian Atlantic area" and play a crucial "role in the marine ecosystem."

Despite this knowledge, the Canadian government, in what now seems like a fit of folly, gave the capelin away. While retaining none for itself, it assigned munificent capelin quotas to other nations, primarily the Soviet Union, and the commercial catch soared from 3,000 metric tonnes taken off Newfoundland in 1970, to 370,000 tonnes in 1976. In 1977 the catch declined; in 1978 it collapsed.

The destruction of major capelin stocks off Newfoundland removed a vital link from the sea's chain of life, and as Scoresby had predicted so long ago, the results were catastrophic. Cod catches of Newfoundland's inshore fishermen declined; whales and seals were presumably affected, although to what extent is not yet known. And, for lack of capelin, their principal food, vast numbers of seabirds are starving to death.

Studies carried out by David Nettleship of the Canadian Wildlife Service in the summer of 1981 showed that at some of Newfoundland's major puffin colonies more than half the chicks died of malnutrition, and even the ones that fledged were so weak, having received only a third of the food they required, that their chances for survival were poor. Murres, another seabird species heavily dependent upon capelin, suffer nearly as badly, and the death of seabirds by starvation is not confined to Newfoundland. At Röst, an island group off northern Norway, nearly half a million puffin chicks have starved to death between 1977 and 1981. In a sea despoiled by man, the adult birds can no longer find the fish to feed their young.

As the more accessible fish stocks are being depleted, exploiting nations, particularly the Soviet Union, Norway, and Japan, look farther north, and south, for new wealth from the sea. The most tempting targets are the Arctic's great stocks of polar cod, and the krill of the Antarctic, the small crustaceans that are the principal food of whales and of millions of seals and penguins. If the mistakes committed in other areas are repeated in the Arctic and Antarctic, the seabirds and sea mammals of these regions, deprived of food, will inevitably decline and may eventually vanish.

Chemical poisoning and overexploitation of marine resources are dangers that loom over the Arctic's future. The danger of oil pollution is imminent. Both the promise and the peril of arctic oil were noted long ago by explorers. Diamond Jenness, ethnologist on Vilhjalmur Stefansson's last expedition (1913-1918) heard from natives on the Alaska coast about a "strange lake of 'pitch' ... which poisoned any bird or animal that drank from it." Now a giant pipeline carries oil from this region across Alaska to tankers that take it to the South.

On his way to the Arctic in 1908, Stefansson travelled along the Athabasca River, in the region where Alberta's tar sands now await an oil price high enough to make their large-scale exploitation profitable. He noticed the tar "which here and there trickled down the cut-banks of the river." Natural gas flared in the wilderness. To Stefansson it was "the torch of Science lighting the way of civilization and economic development to the realms of the unknown North."

He would be amazed and delighted, or perhaps dismayed, if he could see the way in which his vision has been fulfilled. The Beaufort Sea, which Stefansson crossed by dog team in 1914 and where he watched "thousands" of migrating white whales swimming in leads of open water, is now dotted with artificial islands built with gigantic dredges and topped by drills that probe for oil and gas. Immense quantities of natural gas have been discovered in the high Arctic and in a few years enormous LNG (liquefied natural gas) tankers will carry it to the South. It is estimated that by the year 2000, as many as fifteen tankers a day will pass year-round through Lancaster Sound, that seabird- and sea-mammal-rich

region that the explorer William Edward Parry called "the headquarters of the whales." At about the same time, fleets of icebreaker-supertankers, each carrying more than a million barrels of oil, will plough day and night, summer and winter through the ice-choked Northwest Passage with their precious but potentially lethal cargo.

Industry is keenly aware of the dangers inherent in exploration and exploitation on such a gigantic scale in a region as hostile and ecologically delicate as the Arctic. More than a decade ago, T.G. Watmore of Imperial Oil warned that "in the big move north now under way, it [the oil industry] will encounter the most delicate balance of nature anywhere in Canada. We will have to step very carefully, not to do irreparable damage." And Dome Petroleum, Canada's largest oil company, readily admits that "the issue of an uncontrolled oil blowout or massive spill resulting from a tanker accident . . . may be viewed as the single greatest threat to the well-being of the Arctic environment."

Although they promise riches nearly beyond reckoning, the Arctic's oil and gas reserves are finite. They may only last one generation. The life of the seas, unless destroyed, renews itself forever. The care and caution with which the North's mineral resources are developed will determine to a large extent the legacy our time shall leave to future generations: an Arctic blighted and bereft of life, a truly barren land and sea, or the Arctic as it is now, a last great wilderness with a profusion of animals and plants superbly adapted to this inimical region, which the Inuit call *nunassiaq* – the beautiful land.

Circumpolar in distribution, ringed seals are the most numerous of all arctic seals. They are the main prey of the polar bear and formerly were of vital importance to the coastal Inuit.

Bibliography

Augusta, Josef. *Le Livre Des Mammouths*. Paris: Nouvel Office D'Edition, 1966.

Back, George. *Narrative of the Arctic Land Expedition to the Mouth of the Great Fish River and Along the Shores of the Arctic Ocean in the Years 1833, 1834 and 1835*. London: John Murray, 1836. Reprinted Edmonton: M.G. Hurtig Ltd., 1970.

Baird, Patrick D. *The Polar World*. London: Longmans, Green and Co. Ltd., 1964.

Biggar, Henry P. *The Voyages of Jacques Cartier*. Ottawa, 1924.

Bodfish, Hartson H. *Chasing the Bowhead*. Cambridge: Harvard University Press, 1936.

Bodsworth, Fred. *Last of the Curlews*. New York: Dodd, Mead & Company, 1955.

Brown, Dale. *Wild Alaska*. New York: Time-Life Books, 1972.

Bruemmer, Fred. *Encounters with Arctic Animals*. Toronto: McGraw-Hill Ryerson Ltd., 1972.

Bruemmer, Fred. *The Arctic*. Montreal: Optimum Publishing Co., 1974.

Calef, George. *Caribou and the Barren-Lands*. Scarborough, Ontario: Firefly Books Ltd., 1981.

Carson, Rachel L. *The Sea Around Us*. New York: Oxford University Press, 1951.

Clairborne, Robert. *The First Americans*. New York: Time-Life Books, 1973.

Cornwall, Ian. *Ice Ages – Their Nature and Effects*. London: John Baker Ltd., 1970.

Cousteau, Jacques-Yves. *Oasis in Space*. New York: The World Publishing Company, 1972.

Cromie, William J. *Secrets of the Sea*. Pleasantville, New York: The Reader's Digest Association, 1971.

Dagg, Anne I. *Canadian Wildlife and Man*. Toronto: McClelland and Stewart Limited, 1974.

Daniel, Hawthorne, and Minot, Francis. *The Inexhaustible Sea*. New York: Dodd, Mead & Co., 1958.

De Coccola, Raymond, and King, Paul. *Ayorama*. Toronto: Oxford University Press, 1955.

Dunbar, M.J. *Environment and Good Sense*. Englewood Cliffs, N.J.: Prentice-Hall Inc., 1968.

Dyson, James L. *The World of Ice*. New York: Alfred A. Knopf, 1962.

Freuchen, Peter, and Salomonsen, Finn. *The Arctic Year*. New York: Putnam, 1958.

Fuller, William A., and Holmes, John C. *The Life of the Far North*. New York: McGraw-Hill Book Company, 1972.

Hakluyt, Richard. *Voyages*, 8 vols. London: J.M. Dent & Sons Ltd., 1962.

Hall, Charles Francis. *Life with the Esquimaux*. London: Samson, Low, Son, & Marston, 1864. Reprinted Rutland: Charles E. Tuttle Co. Publishers, 1970.

Hanbury, David T. *Sport and Travel in the Northland of Canada*. London: Edward Arnold, 1904.

Hanson, Harold C. *et al. The Geography, Birds, and Mammals of the Perry River Region*. Montreal: The Arctic Institute Of North America. Special Publication No. 3, 1956.

Hoyle, Fred. *Ice*. London: Hutchinson, 1981.

James, Thomas. *The Dangerous Voyage of Capt. Thomas James in His Intended Discovery of a North West Passage into the South Sea.* London: O. Payne, 1740. Reprinted Toronto: Coles Publishing Company, 1973.

Jangaard, P.M. *The Capelin.* Ottawa: Department Of The Environment. Fisheries And Marine Service. Bulletin 186, 1974.

Jenness, Diamond. *Dawn in Arctic Alaska.* Minneapolis: University of Minnesota Press, 1957.

Jensen, Albert C. *Wildlife of the Oceans.* New York: Harry N. Abrams, Inc., 1979.

Kane, Elisha K. *Arctic Exploration: The Second Grinnell Expedition in Search of Sir John Franklin, 1853, '54, '55,* 2 vols. Philadelphia: Childs and Peterson, 1856.

Kane, Elisha K. *The U.S. Grinnell Expedition in Search of Sir John Franklin.* London: Sampson, Low, Son & Co., 1854.

King, Judith E. *Seals of the World.* London: Trustees Of The British Museum (Natural History), 1964.

Kirk, Ruth. *Snow.* New York: William Morrow And Co., Inc., 1978.

Larsen, Thor. *The World of the Polar Bear.* London: The Hamlyn Publishing Group Ltd., 1978.

Livingston, John. *Arctic Oil.* Toronto: Canadian Broadcasting Corporation, 1981.

Mansfield, A.W. *Seals of Arctic and Eastern Canada.* Ottawa: Fisheries Research Board Of Canada. Bulletin No. 137, 1967.

Marsden, Walter. *The Lemming Year.* London: Chatto & Windus, 1964.

Martin, Richard M. *Mammals of the Ocean.* New York: G.P. Putnam's Sons, 1977.

Maxwell, Gavin. *Seals of the World.* London: Constable & Co. Ltd., 1967.

McGhee, Robert. *Canadian Arctic Prehistory.* Toronto: Van Nostrand Reinhold Ltd., 1978.

M'Clintock, Francis L. *The Voyage of the Fox in the Arctic Seas; A Narrative of the Discovery of the Fate of Sir John Franklin and His Companions.* London: John Murray, 1859. Reprinted Edmonton: Hurtig Publishers, 1972.

Nansen, Fridtjof. *Farthest North.* New York: Harper & Brothers, 1897.

National Geographic Society. *Wild Animals of North America.* Washington: National Geographic Society, 1979.

Ogilvie, M.A. *The Winter Birds.* New York: Praeger Publishers, 1976.

Parry, William E. *Journal of a Voyage for the Discovery of a North-West Passage from the Atlantic to the Pacific Performed in the Years 1819-1820 in His Majesty's Ships Hecla and Griper.* London: John Murray, 1821. Reprinted New York: Greenwood Press Publishers, 1968.

Pedersen, Alwin. *Polar Animals.* London: George G. Harrap & Co. Ltd., 1962.

Perry, Richard. *The World of the Polar Bear.* London: University of Washington Press, 1966.

Phillips, Paul C. *The Fur Trade.* Oklahoma: University of Oklahoma Press, 1961.

Prideaux, Tom. *Cro-Magnon Man.* New York: Time-Life Books, 1973.

Pruitt, William O., Jr. *Animals of the North.* New York: Harper & Row, 1967.

Rasmussen, Knud. *Intellectual Culture of the Copper Eskimos: Report of the Fifth Thule Expedition 1921-1924.* Vol. IX. Copenhagen: Gyldendalske Boghandel, Nordisk Forlag, 1932.

Ray, Carleton G., and McCormick-Ray, M.G. *Wildlife of the Polar Regions.* New York: Harry N. Abrams, Inc., 1981.